BEST MOBILE APPS
2013

TOP DEVELOPERS SHARE THEIR SECRETS TO SUCCESS

BY:

JEREMY J. WARNER

Portrait's
PROFILES
OF SUCCESS
2013

ISBN: 978-0-9853555-6-2

Portrait Health
Publishing™

Published by Portrait Health Publishing, Inc.
175 E Hawthorn Parkway, Suite 235
Vernon Hills, IL 60061
www.portraithealthpublishing.com

Original Cartoons by M.Anzal

Table of Contents

Introduction

It took teams of individuals to scour the globe for this year's top mobile apps. When we decided to put the call out for submissions on the best apps of the year, the response was impressive. The way technology has changed our world is evident by the profiles we selected to feature in this year's edition of Portrait's Profiles of Success.

What Is This Book All About?

Many books have been written about great apps and why we need to download them from the App Store. Our team decided to dive in deeper and learn the "behind the scenes" stories about the making of each app and the developers who created them. We interviewed some of the greatest app developers from around the world and they shared the secrets to their successful apps.

We asked the questions to which we wanted answers. What were the setbacks? How did the developers get their app to be #1? And how many downloads are they really getting each day? Many of our featured developers have also provided contact information so you can reach out and share your ideas with them!

Whether you are an aspiring app developer, trying to get your app to the top of the sales charts, or simply love downloading the latest and greatest apps on the market, this book is for you!

Icons Used in This Book

Look for these icons next to each app name to see what's included!

AWARDS – Developers disclose the awards they received for their featured apps

SUCCESS & SALES – Developers share one of their biggest secrets…. their downloads and sales numbers!!!

TIPS & SECRETS – Some of the greatest app developers from around the globe share their secrets to success.

SETBACKS – Not all successful apps rise to the top without a bump in the road. Our featured developers share how they overcame obstacles to become top selling app developers.

"THE MAKING OF" STORY - Top-rated app developers share their inspiration and the stories behind the creation of their stellar apps.

MARKETING TECHNIQUES – Developers share marketing tips and strategies.

RISKS & CHALLENGES – Developers share the risks they took when starting their app business and the challenges they faced while climbing to the top.

PARTNERSHIP OPPORTUNITIES – Now you can share your ideas and app projects with major developers. See which developers want to hear from you!

How To Use This Book

There are five parts to this book and each part represents a different segment of the mobile app market. The individual chapters are divided into app store categories and feature developers from around the world.

We interviewed top developers and included their contributions under their app section. Each key term highlights their answers to the topics outlined above. Some developers have even included a way for you to contact them about your app project!

PART 1: Travel and Transportation

Chapter 1

Travel

"When preparing to travel, lay out all your clothes and all your money. Then take half the clothes and twice the money." ~ Susan Heller

Expedia Hotels & Flights

Free
By Mobiata, LLC.

Description

Save up to 60% with Expedia Mobile Exclusive hotel deals! And now find the best flight to anywhere in the world, too. Flight booking is coming soon to iPad.

Save big on hotel rooms
- Save up to 60% with Expedia Mobile Exclusive hotel deals
- Default to your current location for fast, on-the-go booking
- See reviews from actual hotel customers
- Sort prices, deals or reviews — instantly
- Get cheap hotel rooms or 5-star luxury suites

Find the perfect flight
- Book a flight to anywhere in the world
- Sort by price, duration or time instantly
- Search by airport name, city or code

Book in a flash
- Already signed in? Book in under 30 seconds
- Get Expedia Rewards points for mobile bookings
- Slide to purchase and away you go!

Mobiata

Website:
www.mobiata.com

Email:
info@mobiata.com

Company Profile:
Mobiata creates best-selling mobile travel applications for smartphones and emerging devices. Since its founding in December 2008, Mobiata's applications have been featured by the New York Times, Wall Street Journal, Forbes, Washington Post, TechCrunch, USA Today and in Apple TV and print ads. Mobiata's apps include the best-selling FlightTrack, FlightTrack Pro, FlightBoard and Expedia applications. Mobiata was acquired by Expedia Inc. in 2010 and is headquartered in Ann Arbor, Michigan. For more information, visit the website at www.mobiata.com. Follow us on Twitter: @mobiata.

"The Making Of" Story:
In November 2010, Mobiata was acquired by Expedia, Inc. and now develops Expedia's mobile apps as well as leads the development and research in all emerging platforms for the company. And we still innovate and iterate on Mobiata apps as well.

Successes:
We believe our success comes from a combination of working hard, being meticulous about design, focusing on user experience and making sure that every one of our products is an indispensable tool for travelers. We take exceptional pride in our products, and we think it shows.

FlightTrack Free

Free
By Expedia, Inc

Description
Elegance and simplicity in tracking a flight." — USA Today
NEW! Introducing a fully redesigned and FREE version of the best flight tracking app out there. Download now and see why FlightTrack has millions of fans.

• Real-time status for gates, delays & cancellations
• Zoomable, beautiful flight maps
• Detailed, interactive, retractable flight cards
• Covers 16,000 airports worldwide
• Track 1,400 airlines worldwide

Mobiata

Website:
www.mobiata.com

Email:
info@mobiata.com

"The Making Of" Story:
Mobiata was founded in 2008 when Founder Ben Kazez had an aha! moment while fumbling for various travel documents in the Minneapolis airport. He realized that there should be a mobile app to handle everything he needed while traveling. He set to work and launched FlightTrack for the iPhone in November 2008 and within days it was the best-selling travel app. FlightTrack has remained one of the top-grossing travel apps to this day.

Partnership Opportunities:

We are always looking for talent to help expand our products and features. Please visit our website for current openings. http://www.mobiata.com/careers

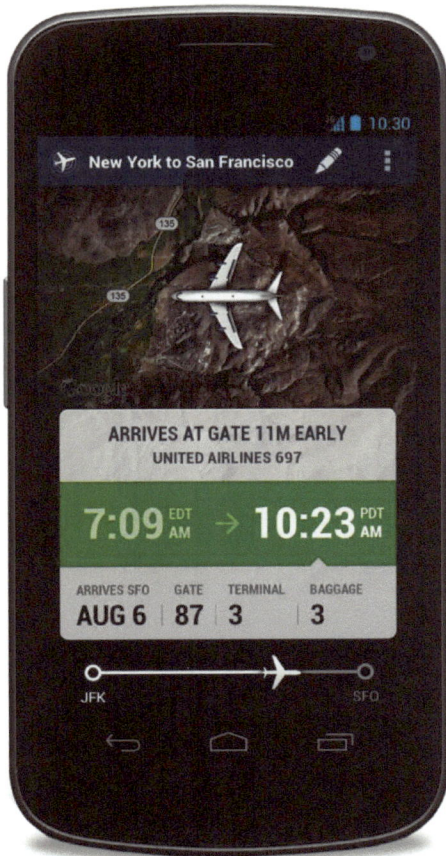

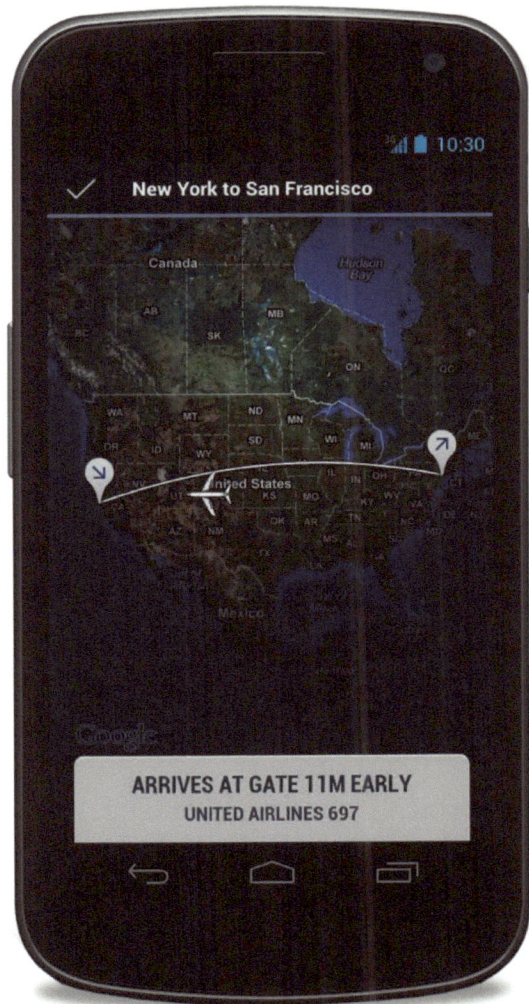

FlightBoard

Live Flight Departure and Arrival Status

$3.99
By Mobiata, LLC.

Description
Winner of the 2012 Webby Award for best travel app!

Turn your iPhone, iPad or iPod touch into the Arrivals and Departures board for any airport in the world! FlightBoard lets you check out the boards of your favorite airports and view all the flight information in real-time. We used the board at Charles de Gaulle Airport in Paris as inspiration for the design so it has a beautiful, old school feel. Download FlightBoard and daydream about your favorite destination.

Mobiata

Website:
www.mobiata.com

Email:
info@mobiata.com

- Covers over 16,000 airports and 1,400 airlines worldwide
- Updated every 5 minutes
- Beautiful interface that mimics Charles de Gaulle Airport in Paris
- See the real-time status for any flight
- Instantly narrow results via search
- Quickly switch between Departures and Arrivals boards
- Easily share flight info with friends and family via Twitter, Facebook & email
- Answer the question: "I wonder what planes are arriving in Venice right now?"

Awards:
FlightBoard received a Webby award in 2012, which was a huge accomplishment for the app. We've had some wonderful accolades and our applications have been featured by the New York Times, Wall Street Journal, Boston Globe, Forbes, Washington Post, TechCrunch, USA Today, Macworld, PC Magazine and in Apple TV and print ads.

New York City Travel Guide...For KIDS!

$3.99
By Go Trexx, Inc.

Description

Kids: this New York City travel app is built just for YOU! This is not your parents' travel guide (but, yeah, it's okay to share it with 'em). With customizable postcards, activities, fun facts, social features and more, this app gives you everything you need to experience the best of the Big Apple, drive family conversations (now YOU can stump your parents with cool trivia challenges), and share your experiences and memories with friends.

Features:

• 13 highlights of the Big Apple: New York City!
• Over 70 pictures gives a sneak peek of attractions
• GPS enabled to alert you when you're near something special
• Kids can quiz parents with "Stump the Folks" questions
• Unlock special features as you visit different places
• Plugged in with Facebook, Twitter, and tumblr

Awards:

Rated as one of the top 50 apps for Parents by Babble.

Making of the Story:

As a step-mom, the Founder and CEO have traveled extensively with children. Whether it was down the street to Grandma's house or on international vacations, she realized how hard it was to find fun apps to help plan the trip and keep kids entertained. Most travel guides require a lot of reading, digesting, and planning.

Go Trexx

Website:
www.gotrexx.com

Phones:
(949) 420-4508

And then most of the time the kids aren't happy with the selections of those planning efforts. So that is when she decided to develop a series of travel guides that allow kids to aid in the vacation planning process and make sightseeing and learning fun. Each app is its own town or travel destination and provides fun facts about that location's top twelve points of interest. There are also cool ways to 'stump the folks' and a customizable postcard feature to enhance the creativity and excitement while on a family vacation.

Tips and Secrets:

Don't tell our young users but the biggest secret about our app is that it is a learning tool. Kids have a great time learning about historical monuments and other sights while having fun and playing trivia games. So shhh!!

Marketing Techniques:

Marketing an app can be tricky. But we have found success by really understanding where parents gather information. We then simply connect with them, share our story, and let the magic happen.

Partnership Opportunities:

Of course you can contact us! We would love to hear from you, and please share any good ideas with us concerning new projects.

Risks and Challenges:

The largest risk was in developing the right technology. There are many development routes (e.g., coding language) and selecting the correct one for a product can be a challenge. There is also some risk in starting an app business right now. There are a lot of players in this space with really good ideas. So it takes time and effort to stand out from the crowd.

Setbacks:

Our company believes in the notion that a setback or disappointment is really a gift wrapped in a different package. These can be hard words to live by, especially in the face of deadlines and revenue demands. But to this day, we have found this notion to be true and have continued to live by it. One 'setback' was that we started to develop our app framework using HTML5 vs. Java or iOS. We realized this decision was the wrong choice and had to start again with new code. But the decision to change and re-do the code made for a better product in the end.

FlightTrack Pro – Live Flight Status Tracker

$9.99
By Mobiata, LLC.

Description
Download now and see how FlightTrack's clean, intuitive design will help you track a flight in seconds. See flight details on beautiful, zoomable maps or get real-time departure info, delays and gate numbers at a glance. Full international coverage means you can track all your flights worldwide. We'll even update you on cancellations and help you find an alternate flight.

FlightTrack Pro works seamlessly with TripIt. Forward your airline confirmation emails to plans@tripit.com, and flights appear automatically in FlightTrack Pro. Works with over 1000 travel sites! It's pretty magical.

TRAVEL SMART.
• Automatically sync itineraries with TripIt
• Sync with your phone's calendar
• Share flight status by email, Facebook or Twitter
• Supports multitasking on iOS 4
• Add flight notes for seat numbers, confirmation numbers and more

BEAUTIFUL MAPS.
• Zoomable, live flight tracker maps with satellite and weather radar imagery
• Review weather radar and forecasts
• Offline mode for use in airplanes – maps still work!
• Airtime, aircraft, and speed & altitude (US)

Mobiata

Website:
www.mobiata.com

Email:
info@mobiata.com

DETAILED COVERAGE.
• Real-time status for gates, delays and cancellations
• Covers more than 16,000 airports worldwide
• Full international flight coverage with 1,400 airlines
• Find alternate flights with a tap

PREDICT FLIGHT ISSUES.
• Push alerts update on flight updates even when app is closed
• Predict flight delays with airport warnings (US airports) and historical delay forecasts

FlightTrack is optimized for crisp, sharp graphics on iPhone 4's high-resolution Retina Display.

Photo Gallery - Hidden Mickey Edition

$0.99
By Flora's Secret, Inc.

Available on the App Store

Description
Discover the Hidden Magic at Disneyland.

What is a Hidden Mickey?
A Hidden Mickey is an image of Mickey Mouse concealed in the design of a Disney attraction or resort. Traditionally, it takes the shape of Mickey's head and ears in silhouette (one large circle with two smaller circles on top), but Hidden Mickeys can also take on many other forms. The most common Hidden Mickey is three circles that form Mickey's head and ears. Other images include a side profile of Mickey's face and head as found on Peter Pan as you fly over the city, a full picture of Mickey Mouse found on Splash Mountain, or even a Mickey stuffed animal hidden on Star Tours. Sometimes, just Mickey's gloves, handprints, shoes or ears appear. Even his name or initials in unusual places are discovered as Hidden Mickeys.

Flora's Secret

Website:
www.florassecret.com

Email:
info@florassecret.com

Features:
- New images added daily of Hidden Mickeys
- High resolution images from real Disney fans
- Take your own pictures of Hidden Mickeys and upload instantly

For just one purchase you will end up with NEW photos every day. You do not need to purchase anything else in order to receive future photos. Photos can be saved, emailed or shared with your friends and family.

Disney Imagineers have been cleverly concealing Hidden Mickeys throughout Walt Disney attractions for decades. Can you find them all? Or maybe be the first to discover and photograph a new Hidden Mickey!

Guide your way through each attraction using our Hidden Mickey photo gallery. Enhance your vacation experience by finding some of the hidden Magic at Disneyland!

Chapter 2

Navigation

"You got to be careful if you don't know where you're going, because you might not get there."
~Yogi Berra

Parkopedia Parking

$1.99
By Parkopedia, Ltd.

Description
The name Parkopedia is a combination of the words parking and encyclopedia (think Wikipedia... but for parking!).

Born out of a frustration of looking for parking, we have set out to map and list every parking space in the world. To date, Parkopedia has grown to cover over 25 million parking spaces in 28 countries around the world thanks to contributions from drivers like you.

Use this app to:
- Find parking using your current location or by entering an address

- Get directions straight to the space

- See parking space availability in real-time (where available)

- Find opening hours, up to date prices, payment methods and more

Parkopedia

Website:
www.parkopedia.com

Email:
christina@parkopedia.com

Company Profile:

Parkopedia.com is the world's leading parking information provider used by millions of drivers and organizations such as Garmin, Toyota, Audi and The AA (UK Automobile Association).

The online encyclopedia allows drivers to find the closest parking to their destination, tells them how much it will cost and whether the space is available.

Parkopedia.com provides detailed information on over 28 million parking spaces in 40 countries, including parking lots, street parking and private driveways and can be accessed online, through SMS or as an iPhone/iPad/Android app and also in in-car satnav devices/smart dashboards.

In addition to providing extensive static information about each car park, Parkopedia also provides real-time space availability information for car parks where such information is available. This allows people to drive straight to an available space instead of circling around looking for parking.

"The Making Of" Story:

Back in late 2007, Parkopedia's Founder and CEO, Eugene Tsyrklevich was in San Francisco attending a tech conference. After driving around the conference center for the 10th time ....he still had not found a parking space.

When he eventually did find a parking space, 30 mins later, he was late and stressed. Back in his hotel room that evening, he started to think about how there must be a better way to find a parking space without wasting so much time driving around....getting stressed and polluting the environment with car fumes. 2 years later, Parkopedia was born and officially launched.

Parkopedia's mission is to be able to answer any parking questions anywhere in the world. To achieve that, we have built parking products that answer those questions wherever the drivers are and that is available online, on mobile and inside a car.

Tips and Secrets:

Work closely with the app stores as they will promote your app for free.

We noticed a huge increase in sales each time we were featured by the iTunes Store as 'App of the Week'.

If you can financially afford to do so, release a 'lite' version of the app (i.e. a free version), to get the early adopters interested as they will spread the word.

Marketing Techniques:

Make sure all links to download the app are prominent on your website, newsletters and social media pages. Try and get a corporate client to integrate the app into their current offering. In Parkopedia's case, this was The AA (UK Automobile Club), the equivalent to the AAA in the USA. The AA wished to develop a new "Find Parking" service to be made available by the AA online and through its mobile platforms.

The aim was for the new service to not only generate additional revenue, but also 'drive' additional traffic to the AA's online and mobile services. In addition, it was hoped that they would also act as a viral customer acquisition tool.

Currently, The AA, uses Parkopedia as its exclusive parking information provider for its AA Route Planner, 84322 SMS service and AA parking mobile apps.

The aim was 'drive' additional traffic to the Automobile Club's online and mobile services. It was hoped that the new parking service would also act as a viral customer acquisition tool to help grow the membership base.

Risks and Challenges:

The main challenge we faced was whether enough people would find the app useful as 'parking' is not really a sexy topic.

We provided a new world/high tech solution to an old world/traditional problem.

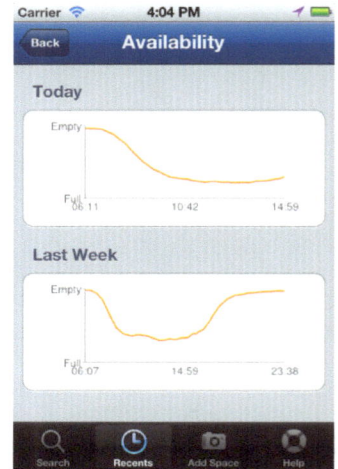

Chapter 3

Transportation

"I failed my Driver's test. Driving teacher: 'What do you do at a red light?' Me: 'I usually respond to texts and check my Facebook.'" ~Anonymous

AA Parking

£1.99
By Parkopedia, Ltd.

Description

Find the cheapest and most convenient car parks in the UK and Ireland using this official AA Parking app. Pay & display street parking is also currently being rolled out across the UK & Ireland.

Benefits:

- Comprehensive coverage of all paid and free car parks in the UK and Ireland
- Save money by finding the cheapest or even free car parks
- Search using your current location or by entering an address
- See parking space availability in real-time (where available)
- Get directions to car park entrance
- Quickly find what you are looking for by using filters such as Free, Covered, Secure and many more.
- Find opening hours and up to date prices
- Find Park & Ride locations
- 24/7 customer support

Successes:

Incorporated in 2009, Parkopedia.com has fast become the one stop destination for all things parking.

Parkopedia

Website:
www.parkopedia.com

Email:
christina@parkopedia.com

Over 1million drivers use Parkopedia's online and mobile products every month.

Over the past 4 years, Parkopedia has collected information on 28 million spaces in 40 countries.

Parkopedia's website traffic has been doubling every 6 months since its launch:

- The Parkopedia iPhone mobile app, in partnership with The AA, was launched in April 2010 in the UK and Ireland. The app was featured by Apple and has been the best-selling UK navigation app since its launch.

- Parkopedia signed the world's largest satnav provider, Garmin, as a client in October 2010.

- The Parkopedia Parking mobile app was launched worldwide in March 2011.

- The Android/ iPad Apps were launched in May 2011.

Awards:
Since its launch in April 2010, the Parkopedia iPhone app has consistently been in the top 10 downloaded navigation apps.

In addition:
- AA Parking app voted as one of the "500 best apps in the world" by The Sunday Times
- AA Parking app Selected as the Best Buy 2011 by Autoexpress in June 2011

Parkopedia nominated for "best innovation in the automotive sector" in the prestigious GSMA Global Mobile Awards 2011.
Parkopedia Apps winner of the Innovation Category of the 2012 British Parking Awards

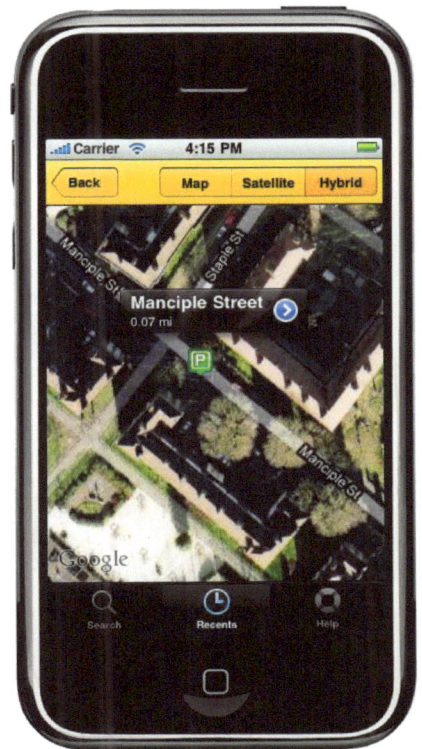

PART 2: Health and Lifestyle

Chapter 4

Food & Drink

"I went to a restaurant that serves 'breakfast at any time'. So I ordered French Toast during the Renaissance." ~ *Steven Wright*

Eat24 Order Food Delivery & Takeout

FREE
By EAT24

Description

Your iPhone can now feed you. No, your phone hasn't learned to cook but you won't have to either, thanks to the free, easy-to-use Eat24 Order Food Delivery & Takeout app.

Order food delivery from over 20,000 restaurants in 850+ cities, with more being added each day. Search from wherever you are, or wherever you're going to be when you're ready for some food. The Eat24 app remembers your preferences, which saves you time, and gives you coupons on a regular basis, which saves you money.

Company Profile:

Enjoy. That's all there is to know about Eat24.

We all worked in the restaurant business for years, so taking care of customers is in our DNA. Unfortunately, it's not in the DNA of the typical delivery-or-takeout experience. So we fixed it.

Having someone cook you exactly what you want and then bring it to exactly where you're sitting is an experience that can't be described in words. Blissful. Euphoric. Exhilarating. Okay, maybe there are a few words that work. The point is, it's freaking awesome.

EAT24
Website:
www.eat24.com
Phone:
1.877.850.4024
Email:
media@eat24.com

But the reality is, most of your time isn't spent dining at fine restaurants. It's spent working your butt off, running errands, and managing life. That's where the Eat24 app comes in. We catapult you to food nirvana whenever you need it – and in just a few taps.

Ordering food by phone presents so many hassles, it can ruin your appetite. Getting put on hold. Crappy cell reception (we're not naming names, carriers). Not being able to communicate with the person on the other end. Reading your credit card digits over the phone. Receiving your order only to find they got it wrong.

These gems and more are why online food delivery is the best thing since sliced bread. And we work to make it the best possible experience for you, with things like:

- Over 20,000 restaurants in 850+ cities – with more being added each day

- Stellar customer support with Live chat, email and phone support – 24 hours a day, every day (literally)

- A service that's 100% free and insanely intuitive to use

- Easy re-ordering and pre-ordering

- Cash back for future orders through CashCoupon

- Exclusive offers and content through our Eat24 VIP program

- Pay with Cash, Credit card, PayPal or the new card scanning feature powered by Card.io

- Reviews, Ratings and Notes for each restaurant and dish

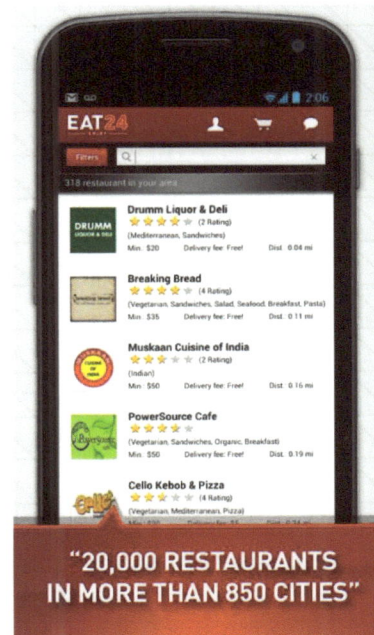

We're here to help you. Enjoy your food, enjoy the time you get back in your day, and most of all, enjoy the delightful feeling of being taken care of. So whether you're working late, in need of a one-night staycation, or just don't feel like cooking – let us handle the food. Your only job is to enjoy every minute of it.

Awards/recognition:
- 8 Food Apps To Chew On -- Forbes
- Item of the Day -- Hello Giggles
- New Eat24 iOS App Puts Online Food Ordering and Delivery In Your Pants -- PandoDaily
- A taste of technology: Five apps for takeout and delivery -- Tech Hive

"The Making Of" Story:
We started developing the Eat24 app as a way to make getting food delivered to you even easier; it's like having a food truck in your pants. No matter if you're on the subway, stuck in traffic, or pantsless on your couch, you can order food directly from your phone. With the Eat24 app, your phone can feed you. Just tap, relax and enjoy.

Successes:
Since its launch last year we've seen an increase in orders and are now getting over half of our business through mobile devices. We've also seen a nearly unanimous positive reaction to our app and our rating is consistently high.

Tips and Secrets:
We do everything to make life easier for our customers. They work hard all day and don't have the time to shop, cook or clean. Our app's success is based on constantly improving the user experience to make your smart phone a straight up genius. This app was designed to let our customers take back a little part of the day to sit back, relax and enjoy.

Marketing Techniques:
We relate to our customers on a human level. When we respond to a customer, whether in customer service, social media, or through email, we talk to them like you'd talk to your friends. Plus we really do love our customers and tell them every chance we can. We're finding that when we give the love out, our customers are more than happy to spread that love with their friends and family.

Partnership Opportunities:
We're always willing to have a conversation with developers. If we can get someone to make tacos instantly teleport from the computer to your home, please contact us immediately!

Risks and Challenges:

One of our biggest challenges was bringing the first class customer service of going to a fine dining restaurant to our mobile app. We worked hard to develop an app that brings you food with the least amount of work. Once you place your order, we take care of everything else. It won't kill zombies, but it will bring you food.

Setbacks:

Every app has its challenges but we work hard to squash all the bugs and make ordering from over 20,000 restaurants in over 850 cities as easy as possible. We are constantly trying new things, adding PayPal functionality, adding Card.io, a cool new way to pay your bill by just scanning your credit card, providing users less time mistyping credit card numbers and more time to look forward to impending deliciousness.

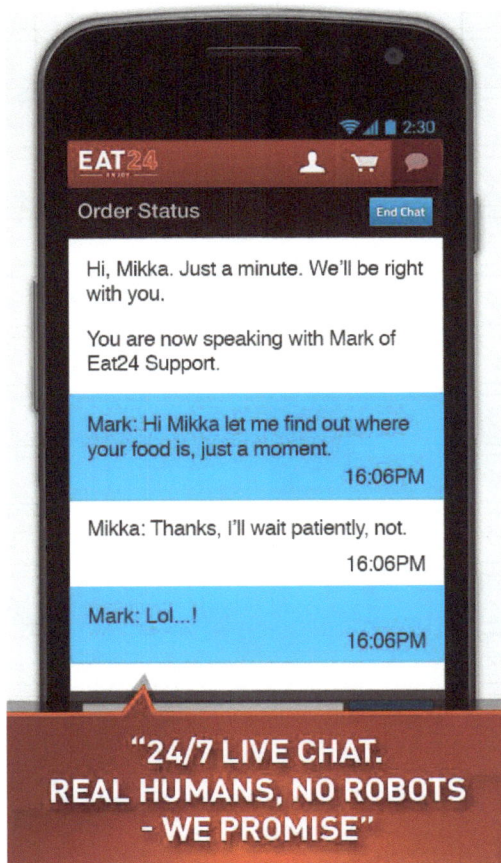

"24/7 LIVE CHAT. REAL HUMANS, NO ROBOTS - WE PROMISE"

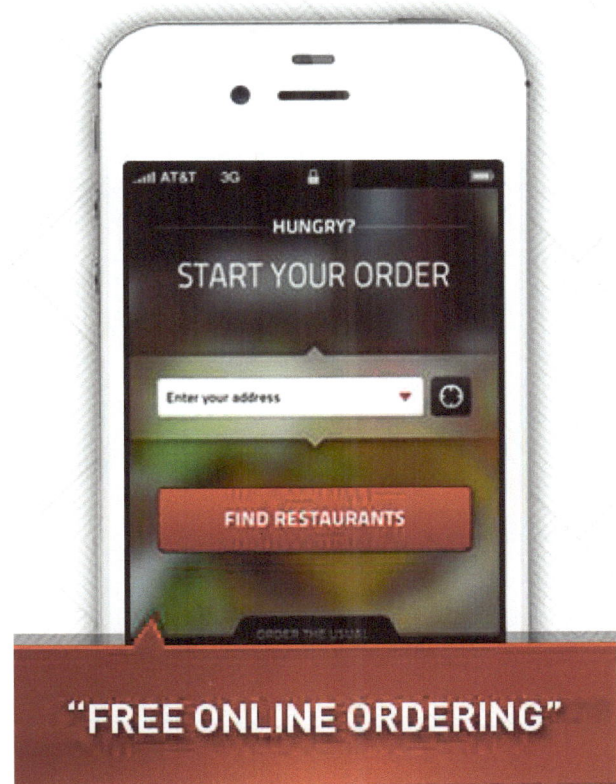

"FREE ONLINE ORDERING"

NOW ACCEPTING **PayPal**

Chapter 5

Health & Fitness

"I really don't think I need buns of steel. I'd be happy with buns of cinnamon." ~ Ellen DeGeneres

Tungz!

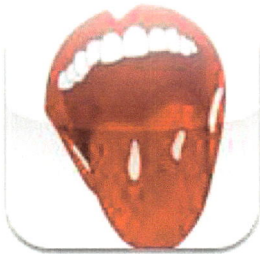

$3.99
By Rohr Acupuncture, LLC

Available on the App Store

Description

Most tongues are created equal...but then life happens! Use this app to discover the health secrets your tongue is telling you.

Based on the thousands of years of clinical practice, tongue assessment is critical to Chinese medicine. By looking at variations of the tongue, including the shape, size, color, and coating, Chinese medicine physicians are able to assess relative health levels and give specific instruction to their patients.

This app is a user friendly guide to look at tongues and see what the variations mean. There are diet, exercise, and lifestyle recommendations based on the tongue variations. Check out your own tongue and be able to assess your friends and family!

Rohr Acupuncture

Website:
www.jamesrohr.com
Phone:
305-753-3399
Email:
james@jamesrohr.com

Company Profile:

James is a Florida licensed Acupuncture Physician and is nationally board-certified in both acupuncture and Chinese herbal therapy. James completed his undergraduate studies in anthropology at Stanford University and his graduate studies in Chinese medicine at the Pacific College of Oriental Medicine in San Diego. He also completed a certification program at Chengdu University in China.

James has extensive history working in integrative centers, providing the best in traditional and modern medicine. He was part of the healthcare teams at Scripps Center for Executive Health in La Jolla, Loyola University Medical Centers in Chicago, and Evanston-Northwestern Healthcare in Illinois. He is currently in private practice and is also the head acupuncturist at a leading integrative center in Miami Beach. Prior to relocating to Florida, James was also a clinical supervisor and professor of acupuncture at Pacific College of Oriental Medicine in Chicago and an instructor in the Health Sciences department for Kaplan University.

His clinical work has been mentioned in the New York Times, Northshore Magazine, and he has appeared doing acupuncture on NBC News South Florida Today. He is the author of numerous articles about health and wellness topics. James is also the developer of two mobile apps: Tungz!, the first mobile application specifically for Chinese medicine tongue assessment, and The Travel Kit, mobile solutions for common ailments affecting road-weary travelers.

"The Making Of" Story:
Tongue assessment is a major component of the Chinese medicine evaluation. I look at tongues all day long in my acupuncture clinic and patients are always asking me what am I looking for? I decided to make an app for that. Everyone has a tongue and now people who have used the app will have a better idea of how to use their tongue as a gauge of their overall health. Look at your tongue today and compare it to the healthy pictures I have in the app. If you look at your tongue again a month from now and it looks further away from healthy, then your tongue is giving you an indication that something is amiss. If your tongue is looking more like the healthy one, then whatever changes you have made are working well.

Successes:
I have been amazed at the response to the app. I did one press release that didn't seem to have any discernible impact. I posted a link on my Facebook account and word began to spread around the Chinese medicine community. To date, I have sold 2,000 apps.

One of the offers I make within the application is I invite users to take pictures of their tongues and email it to me for a complimentary evaluation. I have been inundated with tongue pictures since the release of the app. After one of my evaluations, the client said "Oh my gosh. It is like you read minds also! Amazing that you knew so much about me just based on a picture of my tongue!" I'm proud to have created the very first app exclusively devoted to Chinese medicine tongue diagnosis. When people start looking at their tongues and making healthier, more personalized decisions, then I know developing this app was time well spent.

Tips and Secrets:
I found a topic that I think is fascinating and that I'm an expert in. It's a niche market, those people that want to analyze their tongue, but it has potential wide-spread appeal because everyone has a tongue.

Marketing Techniques:
I targeted my Facebook feed (loaded with acupuncture colleagues) and I reached out to some contacts I had at other professional organizations that have access to students and practitioners of Chinese medicine. I figured if I could get the practitioner to be familiar with the app, then they might share it with their patients.

Risks and Challenges:
My biggest risk was my time investment. I was able to produce the app without too much financial investment. Since I had the knowledge, the biggest initial obstacle was figuring out the coding. Once the app was created, the biggest hurdle was marketing. I decided to do very little initially and see what kind of response I was getting. To have sold 2,000 apps while having spent a total of $150 on marketing, I am pleased with the return.

Setbacks:
When I first had the idea for the app, I spoke with a few developing companies. I was told it would cost anywhere between $10,000-30,000 to make the app. I couldn't believe it! I was a practitioner with a humble practice! I couldn't afford to spend even $5,000 to create an app that might not have any users. So the biggest setback was the first one. I had to figure out how to make my app on a shoestring budget.

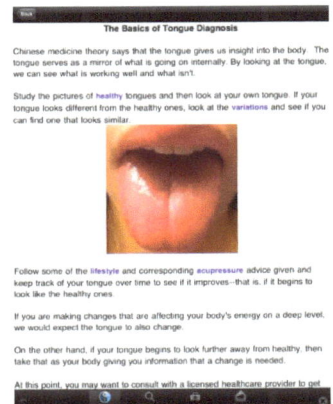

Sociercise – Real Time Running Races

FREE
By Sociercise, LLC

Available on the App Store

Description
Sociercise – Making Every Day a Race Day.

Use the rush of competition to transform your ordinary run into an exciting running race. Sociercise allows you to run and race against people from all over the world in real time. Just sign in, select a race, and run outside. Races occur in real time over real distances with the only real time running races app for iPhone.

Not in the mood to race? You can do an individual workout. Sociercise is also a full featured run tracker, running log, running data, and goal setting system.

Looking to lose weight? Use the motivating power of competition to fuel your running workouts.

Sociercise
Website:
www.sociercise.com
Phone:
(706) 892-6694
Email:
bobby@sociercise.com

Company Profile:
Sociercise was founded in 2011 by Tom Vinkler and Bobby Valentine as a way for people to not only challenge themselves against each other but also against themselves. As the project grew the team realized that the Sociercise platform is a great way for non-profit organizations, charities, and school groups to raise money and awareness for their cause. The goal of our company is to provide a fun way to

exercise socially with friends from around the world while participating in daily workouts and charity events.

Tom Vinkler and Bobby Valentine are proponents of the fit life. Where the combination of exercise, competition, social interaction, and social good combine to form a well-rounded existence. These core principles guide the development of the Sociercise platform. Sociercise is more than an application or a website, it is a way of life.

The story of Sociercise began back in 2009 when the founders first met via an Elance project. Tom Vinkler, living in Hungary, had hired Bobby Valentine in the US to work on a marketing project. Following the project's completion the two entrepreneurs went on with various other projects. Fast forward almost two years and Tom Vinkler again contacted Valentine about the idea for an app that lets people race on a treadmill. From that moment on the two have been partners even though they've yet to meet in person. The entire concept and business of the company has been developed through the use of Skype and email. A truly 21st century business model.

Successes:

The team launched the first version of the Sociercise Real Time Running Races application in the App Store on December 13th, 2012.

Tips and Secrets:

Never underestimate the power of planning for every occurrence. It took us getting testers to realize that certain parts of the registration process, although we did it a hundred times, were flawed. Also, when working with outside vendors both sides need to have complete and up-to-date specs for the project and know each other's roles and responsibilities. Also, don't think that because you build it they will come. There are far fewer App Store Millionaires than there are App Store frustrated and hopeful developers.

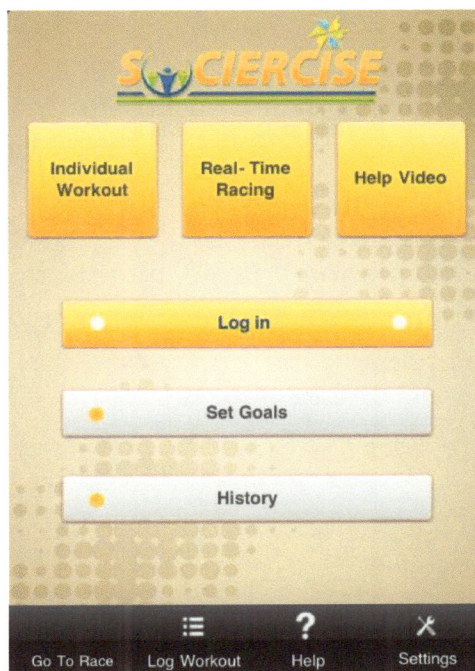

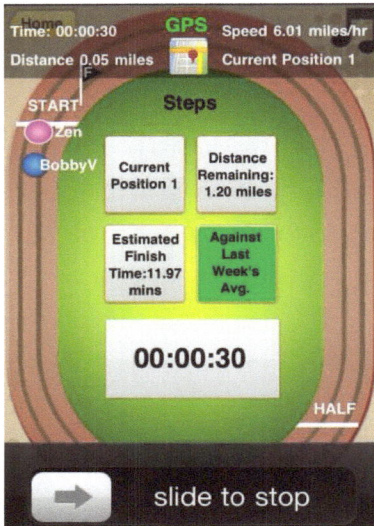

Marketing Techniques:

Sociercise has issued numerous press releases and social media platforms for marketing the app. The team has also entered into contests to promote both the business and the app to users and investors. Marketing gets expensive, especially the traditional formats. We do plan to use traditional media for our Charity Races application.

Risks and Challenges:

By far the biggest risk is money. Both partners in the business have put their own money on the line and the business itself has also taken out several lines of credit. The challenges come from not having enough time, money, or knowledge to get everything done that needs to get accomplished before launching an app. Another challenge some of us older and more mature developers face is managing family responsibilities with development. Like most developers we tend to spend every free minute in front of the computer getting just one more thing accomplished. Taking time for family is necessary and important.

Setbacks:

This ties in along with the risks and challenges. Money is the biggest setback. It takes thousands of dollars to run a successful marketing campaign and after spending the majority of money on designers, specialized developers, testers, etc., there is little left to market the app. The biggest set-back was seeing the planned three month development timeframe turn into almost a year before having a working version in the App Store.

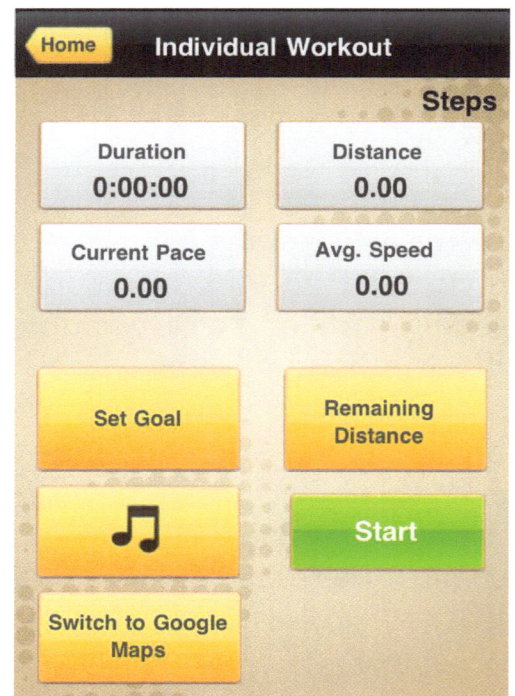

iRideInside

$4.99
By JammyCo

Description

Take your indoor cycling class to go with iRideInside, an app that provides you with "ready-made" as well as "customizable" workouts so that you can get in a great ride anywhere you want and for any amount of time. Choose from a selection of experienced instructors that ride right along with you and push you harder than you ever thought you could go. Whether you are a beginner just wanting to get in shape or an avid cyclist looking to stay fit, iRideInside is the app for you.

Features:
- Video clips of instructors demonstrating the workout.
- Full audio of each instructor's workout.
- Choose the music playlist provided by the instructor or create one from your iTunes library.
- Choose ready-made workouts or customize your own.
- Customize what icons are displayed on the screen, including total elapsed time, rate of perceived exertion (RPE), and cadence.

JammyCo

Website:
www.jammyco.com

Email:
jamima.wolk@jammyco.com

Company Profile:

JammyCo, founded in 2009 by Jamima Wolk, consists of a brother and sister team that produces top-quality, innovative fitness mobile apps. Wolk is a former professional triathlete and triathlon coach and holds a degree in kinesiology as well as an MBA. She started competing as a professional in 2000 and retired in 2007. Her race resume includes everything from sprint to Ironman distance races.

"The Making Of" Story:

I came up with the idea for this app back in 2008 after the birth of my first daughter. I had been taking indoor cycling classes throughout my pregnancy but once she was born I found it difficult to make the class times. I was now on her schedule and found it pretty un-motivating to get on a bike by myself at the gym and try to get in a hard workout. I now have two little girls so I use the app a lot myself at home or at the gym when I can fit in a workout.

Tips and Secrets:

One of the biggest tips I can give is to listen to your customers. If you are getting more than one email asking you to change something about your app, you should take that as constructive criticism and make the change if it makes sense. I also try to answer each email I receive personally within a day or two. Good customer service is key to good reviews and users spreading the word about your app.

Risks and Challenges:

It was a huge risk for me when I started my business back in 2009. I had just graduated from business school and had school loans that needed to be repaid. I had no experience whatsoever in the computer industry but my brother did and agreed to do the programming for my first app, First Time Triathlon. I loved the fact that it would allow me the flexibility to be home with my new baby and luckily it was a success and I have since launched two more apps.

Setbacks:

One of the biggest hurdles I face is competing with larger companies that have a marketing budget. It's easy to get lost in the app store so you have to be a bit more innovative in how you market your app and get the word out.

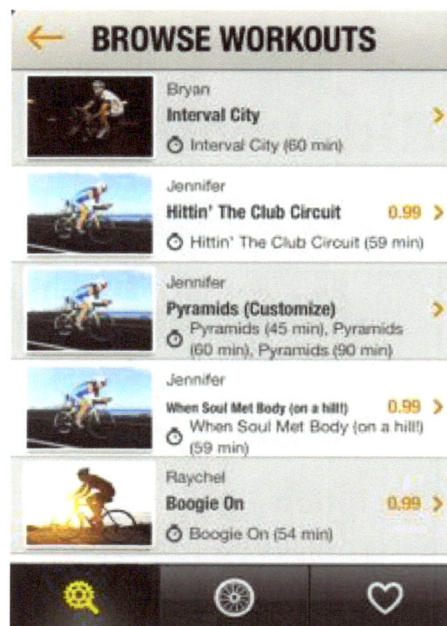

Endomondo Sports Tracker – GPS Track Running Cycling Walking & More

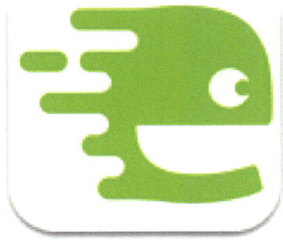

FREE
By Endomondo

Get it at BlackBerry App World

Windows Phone

Available on the App Store

ANDROID APP ON Google play

Description

Make fitness fun with this personal trainer and social fitness partner. Endomondo is ideal for running, cycling, walking and any other distance-based activity.

Join more than 13 million users and start freeing your endorphins!

All data is sent automatically to your profile at www.endomondo.com, where you analyze your training, compete against your friends, follow them live and communicate with the vibrant community of active people throughout the world, no matter what other GPS phone or tracking device they are using. You can also set up automatic sharing to Facebook and Twitter from the website.

Endomondo

Email:
mette@endomondo.com

Company Profile:

The Endomondo Sports Tracker mobile app turns most modern smartphones into a personal trainer and social motivator capable of tracking workouts, analyzing performance, and can aid in the discovery of new routes, activities and insights into fitness so people become and stay active. What makes Endomondo stand apart from other activity tracking apps is its strong focus on the social dimension to fitness. By incorporating aspects found in leading social networks, Endomondo helps users connect with like-minded fitness fans and encourages the sharing of experiences and support for one another in achieving collective goals. Users can send friends real-time pep talks while they exercise, offer valuable route maps, compete against friends for fun, challenge co-workers, and share it all on Facebook, Twitter or across the Endomondo social network.

Awards:

- Microsoft Health Users Group 2012 Innovation Award: Flexible Mobile Workstyle Solutions Category
- Ernst & Young Entrepreneur of the Year 2011: Danish Start-Up
- Handster Best Software Award 2011: Health Software

"The Making Of" Story:

Endomondo was founded in 2007 by three management consultants with backgrounds in sports and athletic activities. Our vision and goal was to make fitness more enjoyable by allowing individual sports to be more social and engaging. We wanted to build a tool that would aid our passion and felt hardware already widely available in smartphones could support the creation of a virtual personal trainer and social motivator for athletes. The vibrant technology community in Copenhagen, along with the personal relationships fostered by the three co-founders over the years, made it possible for us to pull together a creative team that brought our concept to life. In 2008, we unveiled the Endomondo Sports Tracker mobile app that utilizes GPS technology in smartphones to record a full history of workouts for distance-based activities. Additionally, we produced the Endomondo.com website that offers fitness buffs a personal portal to connect to others in the Endomondo community.

Successes:

Now used by almost 14 million people worldwide, Endomondo remains a very popular, highly rated, free mobile app that has helped users track more than 100 million workouts resulting in over 500 million miles of physical activity. To serve the evolving needs of fitness fans, developers are constantly looking at ways to expand Endomondo's service. In 2012, Endomondo integrated with Facebook Timeline, a move that resulted in more than 12 million workouts being shared in just a few months. And in response to user wishes, developers added to Endomondo.com features such as Workout Comparison, Heart Rate Zone Analysis, Interactive Routes, Hydration Guide and a Weather Log.

Tips and Secrets:

There hasn't been any single move or channel that made all the difference in Endomondo's development. It's been a long journey and a lot of things had to come together at the right time for everything to add up to the success being experienced today. However, one very important factor for our ongoing growth has been the app's quality. Endomondo is among the highest rated fitness apps across platforms and that has been a prerequisite for being featured in different app stores and gaining users.

Marketing Techniques:

As with many tech startups, financial resources have been very limited from the beginning and we haven't done any traditional marketing. Instead, Endomondo focused on performing well in the main app stores, on good integration with Facebook and on making a social community where users have incentives to invite their friends.

Risks and Challenges:

Conceptually, Endomondo's co-founders were confident our model was viable and there was a great pool of skilled developers in the region from which to draw. The main challenge, as with any startup, is raising capital. In fact, the founding team ended up working 18 months without any pay to ensure Endomondo's survival.

Setbacks:

All along, Endomondo has intently listened to user feedback and suggestions before we developed and launched new features at our website or on the mobile app. That plan of attack results in not having to make major changes and a user base that feels they're receiving a quality service from Endomondo.

My Walks

$3.99
By Robert Wohnoutka

Description

An Easy to Use, Accurate, Battery-Saving, GPS Walking App for the iPhone

My Walks is a fun and easy app to use whether it's for fitness tracking or just pleasure walking.

It provides advantages over most other GPS walking apps: (1) Easier to use with an auto-pause feature, (2) More accurate with adjustments for uphill climbs, (3) An auto shut-off battery saver, (4) Can get you "Back on Time" or "Back by Dark", and (5) Easy to email your route map and walk history.

My Walks is NOT a pedometer. It does not count steps. It is a GPS-based app that calculates your distance and speed using your change in location. It is designed for outdoor use.

Robert Wohnoutka

Website:
www.wohnoutka.intuitwebsites.com

Email:
rwohnoutka@yahoo.com

EASY TO USE

Just tap "Start" and the app does the rest. No account set-up required. No registration. Doesn't even require a Wi-Fi or cell phone connection, just a GPS signal. Also, you can continue to use your iPhone to make and receive phone calls, send and receive messages and emails, or run other apps.

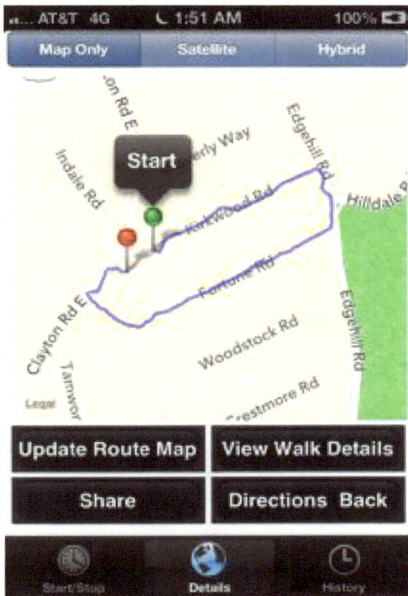

Profile:

Robert Wohnoutka is an independent software developer with over 18 years of software development experience. He is a former Apple employee where he learned the importance of "ease-of-use" which is the first rule he applies to all of the apps he develops. He also has over 20 years of Product Marketing experience with high-tech products where again the importance of "ease-of-use" was his guiding light as he helped companies introduce new technology into the hands of consumers. My Walks is a good example where Robert has applied the "ease of use" rule to an iPhone GPS-based walking application. Unlike other GPS walking apps, this app is very easy to use as it only requires the user to tap "start" to use the app. The app will even end the user's walk automatically should the user forget to tap "End."

"The Making Of" Story:

"My Walks" was Robert Wohnoutka's first GPS-based iPhone app. Ease-of-use was the first rule he needed to apply to make a "Best of App". He decided it should not require any log-on, registration or even a network connection to operate. After all, quite often walkers are in the woods or up in the mountains where network connections are unavailable. The goal was to have a single "Start" button, which would become the "End" button once the app started to get a reliable GPS signal.

Robert soon noticed that sometimes the "End" button would get accidentally tapped and erroneously end the walk. To resolve this issue, he change the "End" button to a double-tap "End" button. Hence, not requiring a "lock" button.

Next, he noticed that he could implement an auto-pause feature, and not require a "Pause/Resume" button. This was very easy to implement. When the user was not moving, they had paused, and when they started to move again, they had resumed their walk! With auto-pause, when the user pauses during their walk, they still got the correct walk duration and average speed.

Next, he observed that sometimes he forgot to end his walk and the information about his walk was incorrect or worse yet, his battery would run down. So, he decided with a GPS signal he could determine when the user had finished their walk, like when the user was in their vehicle and going too fast to be walking, or when the user returned to where they started their walk and was no longer moving much. Hence, Robert implemented Auto-End.

Robert also noticed that sometimes he had a difficult time knowing when to start back to get back before the park closed. He figured, since he knew how far he was from where he started his walk, and what his average speed was, he could calculate when it was time to start back to get back-on-time. So he implemented a Back-On-Time feature and then realized that sometime he just wanted to get back before it was too dark to see. So he figured out how to calculate dusk, and added the Back-By-Dark feature.

One other concern Robert had was that the GPS system did not always give accurate results. So using his math degree, he decided that he could compute which data was statistically out of bounds and disregard the bad data. He could also perform statistical analyses (using least-squares line fitting) on speed and elevation data and compute speeds and elevations that were more accurate.

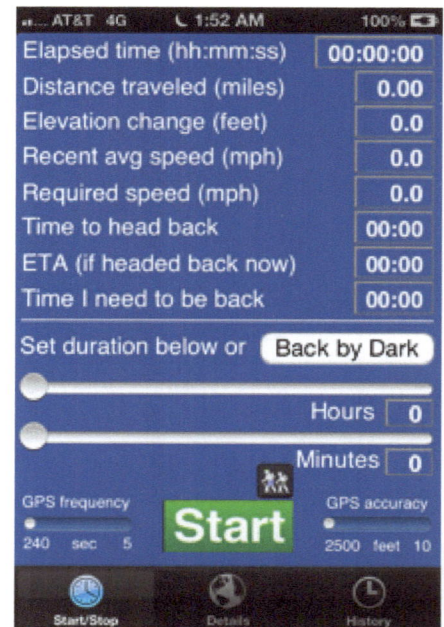

MediSafe family edition medication and pill reminder - Virtual Pillbox

FREE
By MediSafe Project

Description

MediSafe is a beautiful, visual and easy-to-use medication reminder. It will help you take your medicine on time and safely.

As an addition, if you are a caretaker for a parent, child, or patient, you know how stressful it is not knowing for sure if your loved one took their medications on time. If you are on more than one medicine, you know how difficult it is remembering to take each one at the right time.

Medisafe has a solution. It's simple. When it is time to take your medication, the app will remind you. You can also update your app manually. Your caretaker is notified if you don't check in, so they can remind you only if needed.

Reduce your stress and improve the health of your loved one with Medisafe.

MediSafe Project

Website:
www.medisafeproject.com

Email:
info@medisafeproject.com

Company Profile:

MediSafe Project is the first cloud-synced mobile app helping families prevent emergencies caused by over- or under-dosing medications. In addition to reminding users when it's time to take their medication, MediSafe Project sends alerts to selected family members, friends and caretakers when a loved one misses a dose. Aggregated patient behavior data, physician trends and other market aspects are available to help pharmaceutical companies better understand how people receive and take their medications.

"The Making Of" Story:

MediSafe Project was inspired by the accidental and potentially fatal insulin double-dose of brothers Rotem and Omri 'Bob' Shor's diabetic father. When they realized how staggering the problem of medication adherence is – 27,000 preventable deaths per year in the U.S. (one every 19 minutes), making medication non-adherence the #4 cause of death in the US – they developed MediSafe Project as a potential way to lower hospitalization and mortality rates, prolong health through sustainable behavior changes, decrease long-term healthcare costs and help pharmaceutical companies understand patients' barriers to medication compliance.

The company has been active since May 2012, when it was accepted into the first cycle of the new Microsoft Accelerator of Windows Azure program. During the four month intensive program for cloud-based startups, they received mentorship from top entrepreneurs and investors, technical training, and the opportunity to present to angel investors, VC's and media. MediSafe Project was officially founded in August 2012, and had the honor of being selected as Microsoft's BizSpark Startup of the Day in December 2012.

Successes:

15,000 downloads, 300,000 app visits and 250,000 medication doses take in the first 10 weeks of release. Even more gratifying, users' self-reported adherence rate has climbed to 81% - 31% higher than the World Health Organization's average of 50% adherence. For a problem as persistent and far-reaching as medication adherence, a rate improvement this substantial is astounding.

Tips and Secrets:

It's important to research both the problem you're intending to solve and all the competition in the marketplace very well. For us, that meant studying the medication consumption experience – building upon what people liked and overcoming limitations that thwarted earlier efforts.

We invested heavily in a highly differentiated, simple, and beautiful user experience and interface – directing early resources to work with an unparalleled designer.

Marketing Techniques:
We retained a PR agency to help us gain media exposure, but as for pure marketing, we relied solely on word of mouth. Because we solve a universal and stubborn problem – forgetting to take your meds – so effectively, our users have been our best marketers. We're adding a Facebook login option soon, and we plan to leverage the viral potential there.

Partnership Opportunities:
We are actively looking to partner with leading pharmaceutical companies and pharmacy chains.

Risks and Challenges:
The "pillbox app" category was already crowded when we decided to create MediSafe Project. Our decision to leverage cloud technology to sync medication reminder alerts among users' family and friends gave us a clear differentiator and helped us gain traction immediately after release. Our vision extends into a complete ecosystem of patients, families, doctors, drug stores and pharmaceutical companies, and we'll be adding major new elements throughout 2013 – including an automated phone system so users and their loved ones without smartphones can participate, SMS alerts, warnings of hazardous drug interactions, electronic or printable medication lists and dosage logs and more.

Setbacks:
Finding the right sources of funding has not been easy. We're fortunate to continue receiving offers from both VCs and angel investors, but more than once we decided to decline capital because the terms weren't right for us.

Chapter 6

Lifestyle

"When I said that I cleaned my room, I just meant I made a path from the doorway to my bed."
~Anonymous

Intimate Fireplace

FREE
By Game Scorpion Inc.

Available on the App Store

Description
Looking to snuggle up to a warm and cozy fireplace? Then look no further as we have the PERFECT app for you!

Game Scorpion Inc.'s Intimate Fireplace app is one of the best fireplace app's that allows you to display a full size fireplace right on your iPad or iPhone!

The full version comes with over 10 different fireplaces to choose from and even includes various musical backgrounds and authentic fire crackling sound!

To add to it, if you have an iPad 2 or an iPhone 4S with Airplay Mirroring, you can even take the fireplace and put it right on your big screen!

Perfect for special occasions, this app is SURE TO PLEASE!

Company Profile:
Abhinav Gupta is the Lead Developer and CEO of Game Scorpion Inc. He is an App Developer, App Business Trainer and even a Motivational Speaker in the Mobile and Technology Industry. Recognized as a professional in the field, he has had his hand in building and successfully publishing over 25+ apps in over 10+ markets including Apple, Google Play, Amazon, Barnes and Noble, Blackberry, HP WebOS, and several others. He has been developing apps and games for several years and has a passion

Game Scorpion

Website:
www.gamescorpion.com

Email:
info@gamescorpion.com

for creating new things. The apps he and his team have created have been downloaded hundreds of thousands of times across the world in various markets.

Making Of Story:

After paying $10 for a virtual fireplace dvd myself and only getting one looping animated fireplace, I decided that it would be really neat if we created our own app that had over 10 fireplaces to choose from in full 3D. What's even better is that you don't have to spend $10 just to see if you like it, you can download a free copy and if you decide you want more, simply pay $4.99 and get all 10! Not only is it lower in price, it is full of many more options including outdoor fireplaces as well as realistic fire sound!

Tips and Secrets:

We've used the Freemium model which has been really helping this app grow. There are many downloads monthly and there are several methods of marketing we are using to increase downloads. Press releases, contests and word of mouth are really helping to drive downloads.

Risks and Challenges:

I took every last cent I had to start this business and kept at it! Our first app actually took us nearly 8 months to make and thousands of dollars only to yield in its first month live under $300 or so. It was shattering for me at the beginning…I however turned this around and kept focusing on the big picture and now over 25+ apps later in over 10+ markets, I can assure you it was well worth the effort!

LifeKraze

FREE
By LifeKraze

 Available on the App Store

 ANDROID APP ON Google play

Description
Earn points and rewards for your real-life accomplishments!

Here's how it works:
1) Post your accomplishments: Create 160-character posts with the things you achieve throughout the day—big or small.
2) Give Points, Get Points: Each day, you have 300 points to distribute. Give them to the best accomplishments... as chosen by you!
3) Turn in Points for Rewards: Take the Points you earn for your own posts and cash them in for discounts from brands like The North Face, Skullcandy, prAna, O'Neill and many more!
4) Cheer On, Cheer Up: Use High Fives & comments to motivate others. This is an encouraging community. Yep, we're positive.

Follow along with all-stars like Dolvett Quince from NBC's 'The Biggest Loser' and Roman Harper of the New Orleans Saints as they show you how to live like it counts!

Company Profile:
LifeKraze is a new online platform that provides a positive community and rewards for individuals who want to lead active, healthy and fulfilling lives. Founded by college soccer players in the emerging tech hub of Chattanooga, TN, LifeKraze provides a team for everyday life, encouraging social interactions that lead to lasting changes. People share accomplishments, large or

LifeKraze

Website:
www.lifekraze.com

Email:
michael@lifekraze.com

small, through 160-character posts, to which they can attach photos and other links. Every day, each member has 300 Kraze Points to award to other members for their accomplishments. The points people earn can be cashed in for discounts and products from nationally-recognized brands, or converted into charitable donations. With a growing roster of partners including The North Face, Men's Health Magazine, and DonorsChoose, LifeKraze is social media with a mission: To change the way people interact online, to change the way people live offline.

Awards:
- *CNN*: "10 Great Mobile Health Apps"
- *The Guardian*'s "30 Best Android Apps This Week"
- *Entrepreneur Magazine*'s "100 Brilliant Companies of 2011"
- *Greatist's* "The 45 Most Innovative Health and Fitness Startups"
- *Inc. Magazine*: "Finding the Next Big Thing at SXSW"
- *Mashable*'s "Spark of Genius" series
- *WDEF News Channel 12:* "What's Trending – The Summer's Hottest Apps"
- *Appolicious*: Top Android App of the Week

"The Making Of" Story:
LifeKraze is an extension of the team environment that my co-founders and I found as college soccer players. After graduation, we decided to extend the benefits of competition and camaraderie to others, by way of a digital community focused on rewarding and encouraging personal accomplishments. We raised money from angel investors and beta-tested the concept on the web for over a year, before launching our iPhone app. The Android app came six months later.

Successes:
Our successes are the successes of our members. We've seen people quit smoking, undergo major weight loss, complete their first marathon, reconnect with their families, and start volunteering. It's truly amazing what people can achieve with the incentive of Rewards and the force of a supportive community. We're excited to

be able to count amongst these members well-known individuals like Sean Astin (from *Lord of the Rings*, *Rudy*, and *The Goonies*), New Orleans Saints safety Roman Harper, Dolvett Quince (trainer from NBC's *The Biggest Loser*), and more than a dozen Olympians from the 2012 Games. We're also proud of the partnerships we've been able to secure, as a small company: we've worked with The North Face, Universal Music Group, Powerade Zero, Men's Health Magazine, and many other household names.

Tips and Secrets:
If you want a great app, build a great team - other than that, work hard! There are no shortcuts to excellence.

Marketing Techniques:
We've tried to target pre-existing communities, like sports teams, clubs, or classes. People tend to enjoy apps and digital services more if they have friends already on, so bringing groups on together helps for a natural engagement point.

Risks and Challenges:
Entrepreneurship is always risky—I gave up a number of high-paying job opportunities at ad agencies and moved to an entirely new city. One of the challenges we've faced is being based in Chattanooga, TN—it's a wonderful city, but is not a traditional hub for technology startups, so access to industry-relevant talent and resources has been challenging. However, this has afforded us with the opportunity to be creative and develop new approaches that distinguish us as a company and as a brand. We're proud to be part of this community, which is in the midst of reinventing itself to become a center for innovation.

Setbacks:
Because of our early resources and capabilities, we started out with a website, instead of apps. Because we're encouraging people to get up and go do things, mobile has always made more sense, but getting things together to build what we wanted took time. Looking back, though, that progression helped us to learn a lot about the nature of community and how people wanted to use our service—no matter what you do, make sure you're learning and improving!

BuyVia – Price Comparison for Computers, Tech, and more. Black Friday, Cyber Monday Deals

FREE
By **BuyVia**

Description

We help recommend Specific Offers from leading companies such as Amazon.com, AT&T, Best Buy, Dell, HP, J&R, Newegg, Target, TigerDirect, Verizon, Walmart, and others. Tablets, laptops, TV's, headphones, Smartphones, digital cameras, and other more.

Showrooming - Quickly scan Barcodes to find the best price. Set shopping alerts so that you will not miss any price drops.

Access BuyVia anywhere - from your tablet at home, smartphone on the go, then purchase products on your laptop or desktop when you reach home or work.

This App is similar to RedLaser, Price Check by Amazon, the Find, and ShopSavvy, but includes our expert picked Tech deals, local offers and coupons, and product recommendations that are constantly updated.

BuyVia

Website:
www.buyvia.com
Phone:
650-281-5050
Email:
info@buyvia.com

Company Profile:

Based in San Francisco, California, BuyVia is the only online and mobile smart shopping service that helps consumers find quality products at the best price available. BuyVia does this by helping users shop intelligently, allowing them to set their desired price on an item they're looking for, scan UPC/QR codes to see how each store's price stacks up to other stores or online options, automatically get geo-local deals based on where they are, and create personalized wish lists. BuyVia is a combination of hand-curated expertise and custom technology that sorts through millions of products and deals and delivers the product quality consumers will be happy with at the best price.

Awards and Recognition:

- Named one of "The Best Apps for Smart Black Friday Shopping," by _Forbes._
- "BuyVia Is the Must Have Tool for the Upcoming Shopping Season," by _AppAdvice_.
- Named one of the "Holiday shopping essential: Black Friday apps," by _The Washington Post_.
- Named one of "The 10 Best Shopping Apps to Compare Prices," by _PC Magazine_.

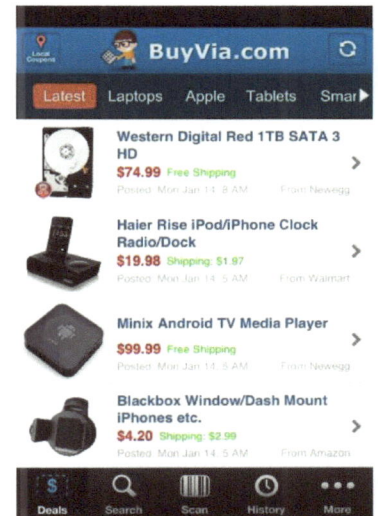

"The Making Of" Story:

As online shopping continues to grow in popularity, the way consumers research and buy has changed as well. With a plethora of tools—the laptop/desktop, the tablet, the smart phone, paper and online coupons, product information and various mobile apps—savvy consumers have both the desire and the means to take virtual and physical "smart shopping" to new levels, allowing them to find high quality products at the best possible price. But more options have resulted in more work and haven't necessarily resulted in smarter buying decisions. With the number of platforms, the tools on each platform (e.g., coupon sites, geo-located "walk by" deals, QR codes, UPC codes, etc.) and the tsunami of information on products that exist, the shopping process is now mired in complexity and has resulted in shopping fatigue.

As self-proclaimed "geeks" BuyVia founders, Norman Fong and Dr. Lawrence Fong, felt this market pain and became dedicated to improving the shopping experience and minimizing deal fatigue by bridging devices and tools, as well as the virtual and physical shopping experiences. The Fongs pioneered the electronic deal space with the founding of TechBargains which was sold to Exponential Interactive in 2007, and Macintosh storage company FWB, sold in 1996 to StreamLogic. Upon launch, BuyVia was primarily focusing in the technology products space based on the founders' past experiences, but they are expanding to other verticals in 2013 to meet the general consumers' needs. BuyVia is more than just a deal website as the product experts have years of experience in product evaluation, price/performance ratios, testing, local deals, and vast knowledge of electronic products, empowering them to cut through the hype and present deals based on an unbiased evaluation. BuyVia delivers the best product/price combination through this curated process, and is the only company to bring together the best of both the online and in-store shopping worlds into one service. For the first time consumers can set a product shopping alert with their desired price so they can be notified when the products they're looking to buy reach their set price, or lower, to avoid missed sales. All of the users' shopping preferences and desired prices are automatically populated across all devices (computers, tablets, smartphones, etc.) via the Cloud so that information is accessible anywhere at any time. BuyVia users can also scan products with the app for a price comparison while they're out shopping, access real-time bargain hunting, and be notified of preferred instant deals based on location. BuyVia is a conversational app-it receives and learns from a person's use and then delivers intelligent results via real time search and/or alerts.

Successes:
As a result of a successful launch from stealth we were positively featured in top tier tech and business publications, including but not limited to *TechCrunch, PandoDaily, Mashable, The New York Times, The Washington Post*, and *Forbes*. As a private company BuyVia doesn't disclose total sales volume and total units sold through the site.

Tips and Secrets:
As consumers we felt the fatigue of searching for the right technology products and the uncertainty of whether or not we are truly getting the best price. There are hundreds of deal sites, price comparison apps, and reviews out there but there isn't anything that streamlines the process, combining all of these features into one, easy-to-use service. It was this realization that drove us to make BuyVia a top selling app. BuyVia brings together the strongest smart shopping arsenal ever available in one app and website, making it easy to find the best products/deals possible across all of the consumers' computers and devices, without tedious and time-consuming research.

We also believe it's extremely important to see how users are interacting with our app to ensure that we are providing the easiest and most integrated shopping experience.

We monitor how consumers are interacting with the app, what features they tend to use, what products they are scanning/searching for and make changes accordingly. For example, since BuyVia launched, we've continued to add additional products/deals based on what users were searching for. This type of ongoing adaptation ensures that we are providing users with a more robust app based on what is important to them.

Marketing Techniques:
We realized that getting positive press was a critical component to our success, so we hired a PR firm that we'd worked with in the past and knew they would deliver results. We first worked with them to develop positioning and spell out the market pain that BuyVia was addressing. Prior to launching our app, our PR firm organized press appointments with key industry leaders where we received valuable feedback that we implemented into our app and messaging development. Throughout launch we continued to watch current market trends while pulling our own data to speak to the press about timely trends and topics, such as price comparison shopping, shopping via mobile devices, "showrooming", and couponing. We conducted shopping and coupon use surveys, analyzed the results, and released the findings to the press which was successful in maintaining momentum with the press. Post-launch we have continued to follow up on current trends, analyze our user data, participate in co-marketing efforts with our partners and demo our app at industry trade shows.

Risks and Challenges:
We were new to the mobile app space and there were many different features to implement with the app. However, we consider ourselves to be experts in the online deal space making this an exciting venture even though it was a new platform.

Setbacks:
When we were in early beta, we experienced mysterious crashes and location-based bugs for the local deal feature requiring us to travel around to replicate. These have been fixed. Now we are continually working on enhancing the app features based on actual user use.

iCall4Help

$2.99
By **The Dreaming Redhead LLC**

Available on the App Store

Description

iCall4Help is fast, simple to use and could be crucial in an emergency. It's an extraordinary and revolutionary way to call for help whenever you need it. And when you can't call for help, the app will do it for you by notify your family, friends, doctor, neighbors or coworkers immediately. You can also share your location via GPS.

With the push of one button you can simultaneously call one person and summon help from unlimited persons via voice messages, text messages and emails.

iCall4Help is customizable to your individual circumstances, unique lifestyle and everyday activities. This will be invaluable to those who have children or elderly parents with medical conditions, those who live alone or travel alone or participate in physical activities such as running, bike riding or hiking. Who benefits? Everyone.

The smart Check On Me feature will verify if you are OK at a specific time selected by you and alert your contacts that you might need help in case you do not respond to the alarm.

The Dreaming Redhead

Website:
www.icall4help.com
Phone:
(805) 895-3903
Email:
support@icall4help.com

"The Making Of" Story:

The iCall4Help app came about as a result of living alone after the death of my husband. I was concerned that there were many possibilities of accidents or mishaps. If I had fallen, how would I get help? There was nothing that I could find that could give me the security I wanted so I decided to create the app myself.

Rather than depend on others to see if I'm OK, I wanted a way to be independent. The Check on Me feature of the app can do this. I can set an alarm and if I don't respond 'I'm OK' the app will automatically alert my contacts. This gives me a great feeling of security and independence.

Risks and Challenges:

The risk was starting a business without really knowing whether the product would be successful. The biggest challenge was learning the app development process as I went along.

Catch Your Cheating Spouse!

A Step-by-Step How To Spy and Phone Tracker Guide

$1.99
By **BustedBooks.com**

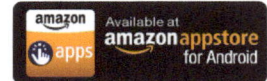

Available on the **App Store**

ANDROID APP ON **Google play**

amazon apps Available at **amazon appstore** for Android

Description

Got a nagging that your partner may be cheating on you....or you're almost certain they are but unable to prove it?

The SPYGuide gives you the software, resources and knowledge you need to get evidence of your partner's infidelity today!

Everything you need to catch a cheater is INCLUDED:

- CELL SPYING
- PC SPYING
- CATCH A CHEATER 101

Don't be a victim any longer....discover the truth today!

Company Profile:

You know what they say about a woman scorned. But it's not always true. Some women will turn their despair and anger over being cheated on, or fearing they are being cheated on, into something better. Something that can help others.

Like Lizbeth Hall; after suspecting her husband of having an affair she needed a resource that could help her find out if he was cheating. And when she couldn't find a one, she decided to write the book, literally. Her mobile app, "Cheating Spouse? How-

BustedBooks.com

Phone:
561-635-5828

Email:
ellie@BustedBooks.com

To Catch a Cheater" is designed to help the suspicious spouse do some serious undercover detective work.

"If you haven't been on the wrong side of an affair, you'll never understand the feeling of despair and frustration you experience. When I finally decided to take action and try to confirm my suspicions about my husband, I became even more aggravated, because there were no real resources to help me. Searching the internet, I was sent on a wild goose chase, spending big money on every software scam and cheating manual I came across. And not one of them actually told me *what* to do; they just gave me a bunch of worthless statistics and useless information," author Lizbeth explains.

Lizbeth Publishing Inc. was founded in 2010 by a stay at home mom who suspected her partner of straying. Prior to divorcing her husband whom she suspected of cheating, Hall tried to find resources available on the internet to assist her in catching a cheating spouse. When none were available, she took it upon herself to create an "all-in-one" resource to help others in the same situation, and BustedBooks.com was born. With sales over $200,000 per year Hall has turned an all too common occurrence into positive cash flow.

Hall believes in the "3E" approach; Educate, Evidence and Encounter. The SpyGuide App first educates the suspicious lover on cheating spouse statistics, and signs that are most common when trying to spot an affair, including how to read verbal and non-verbal deception cues. After completing the "Basics", users move on to the second step; Evidence.

With more than 100 tips, tricks and techniques, users are sure to get the proof they need. The SpyGuide App literally walks the user through, step-by-step, the techniques to gather the proper evidence of an affair. She's even included checklists for the user to follow. This App teaches suspicious spouses everything from how to find and monitor hidden social media accounts, to tracking a partner's movements throughout the day, to how to monitor cell phones and computers for suspicious activity. The final, and what Hall says is the most crucial step, is the Encounter. Confronting a spouse with the gathered evidence is by far the most dangerous step of all. Users have the ability to keep a customized running journal in the app with evidence and proof they've garnered along the way. All in all, it's truly a one-stop shop for learning the truth and tens of thousands of downloads seem to agree.

Awards:

- **FLARE Magazine (Oct/Nov Issue 2012):** App Featured in a Cheating Article
- **Huffington Post:** Named the #2 Creepiest App available in the app store (October 2012)

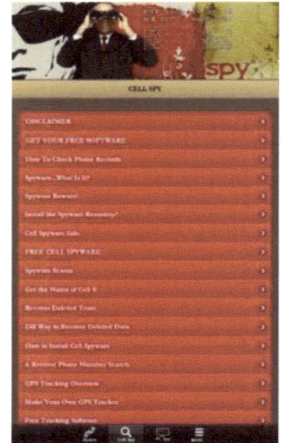

Successes:

TOTAL UNITS SOLD: 160k units (approx)

Tips and Secrets:

When people find out that I'm a one-man operation with a successful app business, they want to know my secret to success. I always tell them that it's not about who can make the best flashlight app or notes app, it's about who can create something that's NOT already in the app store. Creativity is the #1 secret to success in the app store(s). "I probably have three dozen app ideas in development all the time, but you only need one to make it big. So write and submit everything you can think of, sooner or later something's bound to stick."

Secondly, don't sell yourself short. You can learn 100%+ of what you need online. While the mountain that you have to tackle may seem insurmountable at first, if you take it one step at a time you'll be at the top and enjoying the ride down in no time. Learning how to code, becoming a bonafide developer, and managing the process once your app is submitted and approved is documented everywhere on the net.

I read about so many companies that go out and raise $5M to make an app and submit it Apple. To me that's insane! They're already so far behind the eight ball that by the time the app MAY become profitable, they're entirely way too extended. What makes this such a fascinating career is a small time, one-man shop can literally compete with the likes of EA Sports. The playing field has essentially been leveled. This never could happen with a brick and mortar type model.

Marketing Techniques:

The hardest part of being an app developer is marketing. If you're just starting out, I recommend learning how to code and write the app yourself, save your money to hire an internet marketing/SEO expert. I've found that it's not necessary to go out and spend beaucoup bucks. Look to outsource sites like elance.com and freelancer.com. Many bidders on there offer advice just as good as some of the big guys. Believe it or not, I got some of my best advice from a provider on Fiverr.com! Get your feet wet first, don't dive in and bankrupt yourself. You want to get in the green, not go further into the red.

I also recommend marketing to your niche (sounds obvious I know). But, from my experience placing in app ads (i.e. – AdMob, iAds) doesn't give you the kind of bump you need. You'll spend 100x what you earn. Identify your ideal user and learn which websites and publications they read. Place ads where they will see them. Don't market to the masses, your message will get lost. Don't be arrogant and think that your app is better than all the rest so you'll just market to everybody. Here's a tip…it's not gonna happen. Sure there's a 1 in a million chance that you'll get the recognition and become the next ANGRY BIRDS, but you also might win the Powerball.

The best predictor of future behavior that I've seen is the adult porn industry. Seriously, ten years ago you could throw up some XXX pics, make a members only area and charge $20 to every Tom, Dick and Harry. You'd make insane money. However, that era is dead and gone. Similarly, when the app store first opened you could write a silly fart app and you'd be positive in months. Not so anymore, just like the adult industry you have to have a niche.

Risks and Challenges:
The fortunate part of being a single developer is that you don't have a lot of overhead. So, your risk is greatly minimized. You're basically just on the hook for your time, and for whatever you outsource.

Setbacks:
I ran into a few problems initially because I paid thousands of dollars to a company overseas to develop my app. Turns out they couldn't write or spell in English. They basically delivered a bunch of junk code. But, I didn't know how to make an app at that time. I didn't have any money left to hire somebody else, so I started researching how to do it myself. And with a lot of effort and dedication, I eventually taught myself. I outsourced some parts I couldn't figure out and within 60 days I had a complete product ready for market.

Card Lust

$1.99
By **Today I'm Sharing**

Description
Write your heart out. For less than the price of a greeting card, Card Lust lets you send art and love to your friends and family. Illustrations by Elspeth Tremblay and Sara Hingle

What is it?
Card Lust is a delicious greeting app for iPhones. For less than the price of a note card, Card Lust lets you send art and love to your friends and family. For $1.99 you will have 12 decadent designs from Brooklyn-based artist Elspeth Tremblay and another 12 holiday-themed illustrations by Gold Coast-based Sara Hingle. Share your message via text, email or pin to social media. Have a thank-you note in your pocket, a pick-me-up for a pal and never miss a friend's birthday again. Write your heart out, every day.

Today I'm Sharing

Email:
cardlust@gmail.com

How did it come about?
Felicity Loughrey works in a Manhattan creative agency and is the mother of two boys. She always means to send thank you cards and congratulation notes and thinking-of-you missives. But buying a card, writing on it, finding a stamp and mailing it is presently beyond her reach. She looked around for an artful solution and decided to make her own.

Marketing Strategies:

Card Lust is a lifestyle app targeted at women. I like to think that it has been successful as it has a very distinctive look and is unlike any other app in the App Store. The illustrations are hand-drawn so they are very special.

As Card Lust is such a girly app our main marketing approach has been blog and influencer outreach. For most bloggers and influencers we offer to gift the app and let them try it. And we've had a tremendous response with being included in best-of lists and featured on blogs that we greatly admire.

Greetme

FREE
By Greetme

Description
Greetme is the new way to let your friends know just how important they really are; It allows you to send, receive and store memorable eCards.

Cards are currently available in five categories: Birthday, Love, Get Well, Wedding and Christmas. New categories are being rolled out regularly, and the collection of cards is expanding on a weekly basis.

Sending a card:
1. Simply select a greeting card category.
2. The app's seamless integration with Facebook allows you to select a recipient.
3. Choose the card which best fits your loved one's personality.
4. Type in a personal message, and you're done!

Greetme

Website
www.greetme.com

Email:
greetmeapp@gmail.com

"The Making Of" Story:
The app actually started as a side project portfolio enhancer. Right from the beginning the feedback had been amazing, and more resources were added to it to improve and make it better.

Successes:

On its first week in the App store, the app was downloaded to over 600 devices. By the end of the first month 4000 users had been using Greetme to send and receive eCards.

Tips and Secrets:

I think there 2 reasons people respond well to the app. First, because the content doesn't feel cheap, like most free alternatives in the market. Second, the app is all about what the app actually does… meaning it's all about the cards and the personal message. There's no clutter and even on our web site, our logos, marketing statements, etc., take a back sit to the cards in terms of design and UX.

Risks and Challenges:

We have launched the app with a next-to-nothing budget for Facebook ads. Because the app is social in its nature, each card sent also helped reaching more users.

Saving Memories Forever

FREE
By **Saving Memories Forever, Inc.**

Description

Create your family history through audio recordings with the help of this special iPhone App from Saving Memories Forever.

Have a story to tell? Wish to share it with current and future generations? Tell it in your own words, exactly as you remember it, and use this App to upload it to the Saving Memories Forever website.

The App and website will allow you to record as many stories as you like, then share them with whomever you wish, as you create your own audio family album.

You can talk about the first time you met your spouse, the birth of your children, a letter your grandma wrote you, or your favorite vacation. Or you can talk about sports achievements, first cars, and college days. It's your choice.

Saving Memories Forever

Website:
www.savingmemoriesforever.com

Phone:
314 329-8496

Company Profile:

Saving Memories Forever is a unique service that allows anyone to create an audio biography and share it with his or her family. The iPhone and Android app quickly and easily leads you through the process. The app suggests questions, records the story and seamlessly uploads to the Saving Memories Forever website (www.savingmemoriesforever.com) where they can be shared with relatives. The site is private, only people that have been invited can listen to the stories. Each recorded story can have up to 20 text or photos attached to it. You can tell a story about a vacation and have pictures showing where you were. You can record a story about a recipe and attach text showing all the ingredients. In addition, you can tag your stories to provide another layer of organization.

Saving Memories Forever, Inc. was co-founded by Harvey and Jane Baker. Harvey is a former President and COO of several companies and Jane is a marketing public relations professional as well as a teacher. Harvey and Jane were interested in their family history and twenty years ago they researched and had an artist draw a family tree. As soon as it was done, they realized that although they had the facts of their relative's lives, they had no idea of the person behind the names, dates and places on the family tree. They didn't know what their relatives thought was important or interesting or funny. Harvey and Jane started on a quest to find a way to truly know the person behind the name on the family tree and developed Saving Memories Forever, where the small stories that make up a rich life can be recorded and shared with relatives for generations.

Awards:
First Place Project Innovation from Springfield Area Chamber of Commerce, Springfield Illinois 2012. St. Louis County Economic Counsel Semi-Finalist, St. Louis, Missouri 2012.

"The Making Of" Story:
There were three events that led to the inspiration for Saving Memories Forever. The first was an interest in our family history, which led us to produce a family tree about twenty years ago. The second came from my cousin who was recording his father, my uncle, before he passed away. My cousin told us about his recordings, but we knew that we would never hear them. They were too long and there was no organization to them. The third was the advent of the smart phone with the ability to write apps that utilized the built-in recording and file uploading capability. From the family tree we realized that we really wanted to know more about the day-to-day life of the relatives on the chart. We wanted to know their thoughts, what made them happy and sad and anything else they thought was important. We wanted to understand them in a deeper way than the chart would allow us. From the recordings of my uncle we realized that this was a way to get a rounded understanding of a person. There is something rich and interesting when listening to a person tell a story. The inflection

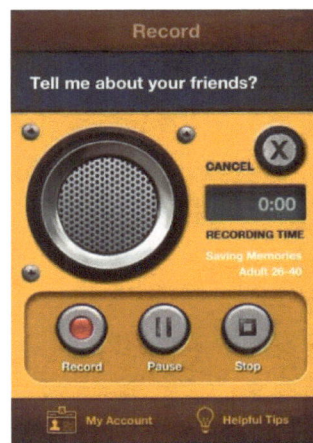

and intonation tell you so much more about the person than words on a piece of paper. We also realized how easy it was to record a voice, how little preparation was needed. We supplied the organization through the questions we suggest. The invention of the smart phone was the technology that unlocked the ability to record and share family memories quickly and easily. With apps, we were able to create a simple easy-to-use system that provided questions to answer, a simple recording interface and an automatic upload capability at a very affordable price.

Successes:

We have over 400 subscribers on the site. We have received very positive reviews from a number of bloggers and magazines. We have had several newspaper articles including one in The Saturday Evening Post and in the Andrews Gazette, Andrews Air Force base paper.

Tips and Secrets:

In order to advertise our app in the marketplace we are using a variety of social and traditional media. On the social media side we have a Facebook page, a twitter feed, a blog, a YouTube account and a business Pinterest account. On the traditional media side we have articles and press releases in a variety of publications. We have also done interviews for radio stations, some of which became podcasts. We have identified our markets as genealogy, reunions, parenting and seniors. We are working in each of these markets to provide an understanding and appreciation for our product. Where else can you quickly and easily create an audio biography, share it with your family and leave a living legacy for generations?

Marketing Techniques:

In each of our target markets we have developed a message that we use. In genealogy we talk about the power of the voice and the ease of use. The genealogy market is made up of people who research and document their family histories. What could be better than being able to hear the voices of relatives telling stories about what they thought was important? In the reunion market we talk about sharing the moment and the stories of the family. We have developed a technique we call "pass the phone" where each person in the family will record the answer to a particular question like, Where did you grow up? This is the essence of the family reunion, a place for people to re-connect and share the memories of their lives. Saving Memories Forever provides a quick and easy way to capture the stories and share them with others, some of whom were not there. In the parenting market we talk about recording the stories about your children as they grow up and having the children record the stories of their relatives using the system. In years to come these recordings will be treasured. Think about having the stories about your child's childhood when that child grows up and gets married or is having their first born. A special market within parenting is the military market. We are offering our premium service to active military for free. We believe that giving our military families the ability to record bedtime stories when they are deployed overseas and have the spouse play them back at bedtime for their children is a great use of our system. Knowing those stories will be around for generations makes answering the future question, "What was my grandfather like?" as quick and easy as going to the computer and playing back one of the recorded stories. In the senior market we talk about how easy and fun it is to create a living legacy. Seniors are interested in what the family will remember about them. What would be better than creating a living legacy now?

Partnership Opportunities:
We are open to discussing extensions to our app. We would be happy to discuss our app and its capability with others. Anything that extends the usefulness of our app to our target markets is open for discussion.

Risks and Challenges:
Whenever you start a new business there are risks and opportunities. The first risk we had was making sure that the idea was practical. We used industry standard software to minimize that risk. We put together existing proven elements in a new and different way to create a unique service. The second risk is the market. Would the idea catch on? We did focus groups before we launched that gave us some assurance that the idea was viable. The final risk was the financial risk of working on a new business. Fortunately we had money saved for this type of risk and were able to get a bank loan and a line of credit to finance the business.

Setbacks:
We had hoped to provide an automatic transcription capability for the stories that were recorded. Unfortunately we felt the current state-of-the-art for translating voice to text was not up to our standards and we had to shelf the idea. However, it still lives in the application where a special file is available for all stories called transcription. For now, people can transcribe the story themselves and upload it to the site. In the future, assuming the voice to text capability improves to the point we can use it, we have the capability of re-integrating transcription into our system.

Coupon Sherpa Mobile Coupons

FREE
By **The Frugals**

Description

Of course we think the Coupon Sherpa app is great, but don't just take our word for it. Kiplinger loved its convenience and money-saving power calling Coupon Sherpa, "One of our favorite free iPhone and iPod touch applications." The same sentiments have been echoed by major media and retail experts, with rave reviews from numerous media outlets across the country. The app was even voted an iTunes Staff Favorite just one month after its release.

Thanks to Coupon Sherpa's mobile coupons, the days of clipping them from the newspaper are indeed over. Now, no matter where you go (as long as you have cell service, of course), you have access to hundreds of coupons for some of the most popular stores around.

Simply find the store you're shopping at in the ever-expanding retail section. When you locate the deal you're looking for, show the cashier and they'll scan in the savings directly from your phone. No printing necessary. The mobile coupons are updated daily so you always have access to the latest discounts and special offers as soon as they become available.

Whether you're a shopaholic or only visit the mall out of necessity, you won't have the option for extra discounts unless you keep it installed on your iPhone or iPad.

The Frugals

Website:
www.couponsherpa.com
Phone:
(970) 672-1136
Email:
info@couponsherpa.com

Company Profile:
The Coupon Sherpa app is the first-ever mobile coupon app offering digital coupons for hundreds of retailers, restaurants, movie theaters and more. Created by a small business in Northern Colorado, the Coupon Sherpa mobile app has been downloaded over one million times and praised in such media outlets as Wired, PCWorld, Good Morning America, The New York Times and many others. The app is available for free on both the iPhone and Android smartphone devices.

"The Making Of" Story:
Internet entrepreneur Luke Knowles is not only committed to helping people save money while shopping; he's also an advocate for eco-frugal living. After the popularity of his first website FreeShipping.org -- a site dedicated to offering free shipping codes and online coupon codes for online shoppers -- he wanted to create something that helped in-store shoppers redeem coupons digitally.

Since the smartphone was becoming a regular accessory for shoppers, Luke came up with the idea for an app that enabled consumers to access discounts from their phone. Luke knew that being first to the market with this concept was crucial to the app's positioning, and quickly got to work developing and testing the app for the iPhone. In March 2009, the Coupon Sherpa app was accepted by iTunes and within one month was selected as an iTunes Staff Favorite. In 2011, the Luke's team developed an Android version in response to consumer demand.

Successes:
March 2009 -- Coupon Sherpa becomes the first mobile coupon app released to the market.
April 2009 -- Coupon Sherpa is named an iTunes Staff Favorite
October 2012 -- Coupon Sherpa is featured in iTunes Top 100 Free Apps List and Top 10 Free Lifestyle Apps

Tips and Secrets:
Part of our success can be attributed to being first to the market with a mobile coupon app. Being first at anything gives you an advantage but the product has to be good in

order to maintain media and user interest. We continue to look for ways to improve the app and include users in the process by making ourselves available for comments via social media outlets.

Marketing Techniques:

When we released the app in 2009, the response was overwhelming. The app was selected as an iTunes Staff Favorite within one month and we decided to create a website that not only promoted mobile coupons, but also offered online and printable coupons, too. We wrote and promoted blog posts about ways to save money and positioned ourselves as experts in frugality. In turn, media came to us for suggestions.

Risks and Challenges:

While there's an advantage to being first to the market with something, there's also the inherent risk of it never being done before. There were plenty of unknowns and "what if's" that we had to work out, plus all the research and testing was up to us. The biggest challenge was testing the app at retailers and getting stores on board with the concept of digital coupons. This practice is more mainstream now but at the time, it felt pretty daunting to hand a smartphone to a cashier in order to redeem coupon savings.

Setbacks:

In addition to restaurants and retailers, we wanted our app to include grocery coupons since it's the most popular type of discount. Though the app can access mobile grocery coupons, grocery stores require a paper trail of manufacturer's coupons and as such, our app is often not accepted by grocers. However, shoppers can use the app to upload available grocery coupons to their supermarket loyalty cards.

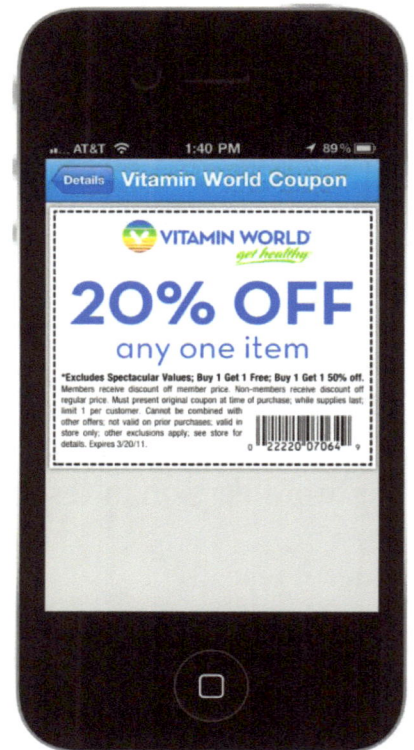

Clutch Shopping

FREE
By **Clutch**

Available on the App Store

Description

You want to use your mobile phone to shop your favorite merchants and products? You're too busy to search for the best deal that is right for you, your family or friends, or fumble through 10 apps to make a purchase or send a gift? You want one app that has all the shopping features you need to be an efficient and effective shopper, using only your mobile phone.

That app is here!

Clutch is the only mobile app where you can get the best deals, give the best gifts, and share these finds with your friends and family in a single place.

Clutch combines everything you need to shop with your phone – mobile wallet, coupons, daily deals, shopping searches, shopping lists, price comparisons and gifting – with the ability to make purchases and deliver them anywhere, with a few taps.

Clutch

Website:
www.clutch.com

Email:
contact@clutch.com

Company Profile:

Our mission at Clutch is to simplify the mobile shopping experience. We understand that keeping track of gift cards, coupons and loyalty cards can be cumbersome. With Clutch, consumers can store all their cards in one place on their mobile device. Search for products – both online and locally – to find the best prices and save even more time by reviewing and purchasing daily deals from more than 20 services right in the app. No more time wasted checking 10 different apps.

With Clutch, looking for gifts has never been easier. Consumers can import friends' birthdays from Facebook, save the brand they like to their Clutch List, then see the latest coupons, offers and gift cards each time they open the Clutch app.

Spend a bit more time with Clutch and you'll find even more ways it can simplify – and improve – the way you shop with any mobile device.

For merchants, Clutch provides the only mobile platform that unites shopping, loyalty and gifting, delivering the most relevant offers to consumers and the most targeted customers back to the merchants.

Based outside of Philadelphia, the Clutch team is composed of executives who have played vital roles in a number of successful entrepreneurial ventures and worked with some of the largest brands including Coca-Cola, HONEST Tea, eBay and Nike. Clutch is privately held and has secured an undisclosed amount of funding.

"The Making Of" Story:

My co-founders are serial entrepreneurs with experience in retail (Half.com and eBay) and with big brands like Coca-Cola, Honest Tea, and Nike. We started building the technology four years ago because we saw that merchants, especially tier 2's, would need a mobile commerce platform to get to market in a much more cost-effective way than having to build it on their own. We formally started the company in early 2012 and launched the consumer app on December 6, 2012.

We wanted to use a name that was simple to remember, hip, relevant and gender neutral. We settled on Clutch for a couple of reasons.

1. A "CLUTCH" is a woman's small purse, and because our app is used for shopping/gifting/offers and because we are effectively "downsizing" some of the need for woman's clutch by allowing her to virtualize cards, coupons. etc., we believe this is an idea that women would connect with.

2. CLUTCH is defined in the urban dictionary as being "just what you need, just when you need it", which is what CLUTCH was designed to be from a content and convenience perspective. This is a concept that allowed for a lot of flexibility around branding that could tie to more traditional male thought

processes. As a simple example of this, a sports player is considered "good in the CLUTCH" or a "CLUTCH player" because they deliver when the game is on the line. They give you what you need, when you need it.

Successes:

Since its launch on December 6, 2012, Clutch has been downloaded by over 20,000 consumers. In addition, Clutch currently supports hundreds of leading brands and merchants and aggregates more than 100,000 deals, offers and rebates daily.

Tips and Secrets:

In short, Clutch spent four years understanding the pain points consumers would have when using mobile shopping apps and designed our app to give them what makes their lives easier. Clutch is the only mobile app where consumers can get the best deals, give the best gifts, and share it all with friends and family in a single place. Its proprietary recommendation engine provides users with highly relevant and timely shopping offers, tailored for them based on time, location and personal data including gift lists, loyalty programs and offers. Consumers can also make purchases within the app, which integrates with Apple's Passbook, with just a few taps.

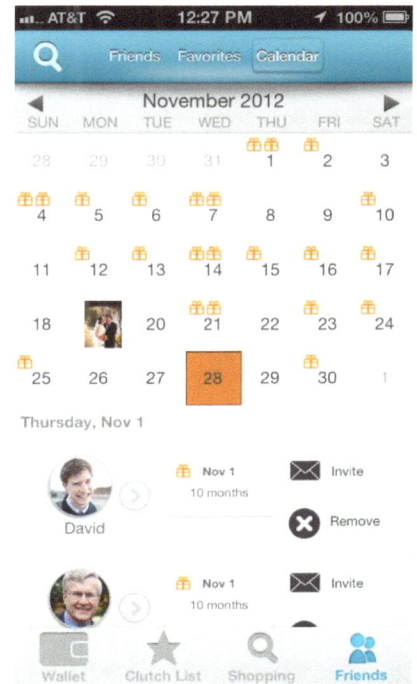

In addition, gift giving is made simpler and more intimate with Clutch. Integration with Facebook enables users to quickly access their friends' wish lists for smarter gift purchasing decisions. When purchasing an eGift card through the app, consumers can personalize it with a video, photo and/or message for each gift recipient, making the experience less transactional.

Marketing Techniques:

Our marketing strategy consists of two dimensions. In order to drive downloads of the app among consumers, we are leveraging PR to drive coverage of the app and gain app reviews. We also are running promotions with online communities most popular with our primary user demographic: busy moms of all ages who are responsible for 85% of all shopping decisions. Other tactics include sweepstakes and giveaways, affiliate marketing channels, partnership-driven distribution and a Clutch-based loyalty and rewards program to drive activity and downloads.

The other dimension is B2B marketing of our "private-label" platform. For this, we are targeting what we call Tier II merchants — retailers with 50 to 750 regional locations — that do not have the resources or inclination to build their own mobile platforms.

Partnership Opportunities:

Yes. We can support merchants either though integrating their brand, coupons, offers, gift cards, merchandise inventory, loyalty programs and more into our direct-to-consumer platform (Clutch) or we can provide a customized "Private Label", proprietary application using our complex suite of partnerships and industry integrations. We can typically execute a completely integrated white label solution within a 6-8 week time frame.

Risks and Challenges:

I gave up an income during a sagging economy to start this company. While building the business, I got married, and had two little girls. I sacrificed so much and so did my family to chase this vision. It's the ultimate leap of faith when you have to risk everything in the hope that in the end, someone likes what you've done.

The challenges were constant in the beginning because the infrastructure we've built, through the complex system of partnerships we've fostered, didn't exist. We had to build it as we went and once we had the partnerships, it had to be designed into the product in such a way that it was clean, intuitive and meaningful to the users.

Setbacks:

Everything we've accomplished has been through trial and error. We learn quickly but the fact remains, success is earned on the back of countless and sometimes painful failures. But, there was a point at which we realized that if we didn't quit, we'd succeed.

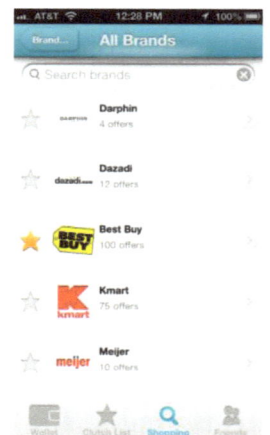

Chapter 7

Tools

"A computer once beat me at chess, but it was no match for me at kick boxing." ~ *Emo Philips*

Linquet Anti Loss & Anti Theft

FREE
By Linquet Technologies Inc

Description

NEVER LOSE ANYTHING AGAIN! Linquet is the easiest Anti Loss and Anti theft solution for your phone and valuables like keys, wallet, tablet, laptop, backpack, sunglasses, etc. Install this FREE app to track your phone. You can also get your Linquets from our site (Linquet.com) to STOP losing your stuff!

We all lose/misplace our phone, keys, bag, tablet, pet, wallet, purse, laptop, etc. Losing your phone or valuables could be anything from upsetting (e.g. going crazy to find your phone) to potentially devastating (e.g. being a victim of identity-theft).

Linquet is the only anti loss and anti theft solution which truly PREVENTS you from losing or misplacing your phone and other stuff.

Linquet Technologies

Website:
https://linquet.com

Phone:
604-779-9140

Email:
info@linquet.com

The Making of Story:

We're all losers! In the US alone, more than 160,000 phones are misplaced, lost or stolen every day, costing Americans $30 billion a year! And the number of lost or stolen valuables is even higher than that. We all have busy lives and we constantly forget/lose our phone, keys, bag, wallet, purse, laptop, etc. Losing your phone or valuables could be anything from upsetting (e.g. being late for a meeting) to potentially devastating (e.g. being a victim of identity-theft). Linquet is the result of our long-time obsession with solving the simple yet huge problem of "losing/misplacing our phones and valuables".

Successes:

During the short time we've introduced Linquet, we have gotten praised by the press, have raised money from international investors and have shipped Linquets to 5 continents. Much more to come, so stay tuned!

Tips and Secrets:

UX is King! User Experience and the way you solve the user's problem is the most important thing in designing and building a great product.

Marketing Techniques:

Think about, empathize with and love your users.

Risks and Challenges:

We had many different aspects to our solution. As opposed to just an app, we had to both design and develop a cloud-based hardware-software solution that would just work. Just the supply and manufacturing aspect of our product was a startup on its own.

Setbacks:

We initially thought that we could outsource the hardware design. But, as we learned the hard way, it's impossible to develop a very innovative product without designing and controlling every single aspect of it.

PART 3: Business and Productivity

Chapter 8

Business

"If at first you don't succeed, skydiving is not for you."~ Anonymous

Adobe EchoSign

FREE
By Adobe

Description

EchoSign from Adobe, the #1 eSignature service lets you get documents signed and access your EchoSign account directly from your iPad or iPhone.

- Send documents for legally binding eSignatures and get them signed in minutes rather than days.
- Get documents signed instantly on your iPad or iPhone when meeting a signer in person.
- Send documents from your EchoSign library, Photo album, email attachment or from other Apps such as Dropbox, Box, etc.
- Track the status of your agreements with real-time status updates.
- View signed agreements stored in your EchoSign account

Sign in to your EchoSign account or create a FREE EchoSign account and start using the Adobe EchoSign App immediately.

The way the world signs.

Adobe

Website:
www.echosign.adobe.com
Phone:
1.877.324.6744

With over 4,000,000 users and 50,000 customers including Google, Facebook, Groupon, Living Social, Twitter, Symantec, VMware and many more, Adobe EchoSign is the leader in Electronic Signatures.

Legal, Trusted

Fully compliant with the Federal ESIGN Act, UETA, and the Electronic Commerce Directive, EchoSign provides protection for both the sender and the signer during the signing process including key authentication and privacy, fraud protection, and consumer disclosure.

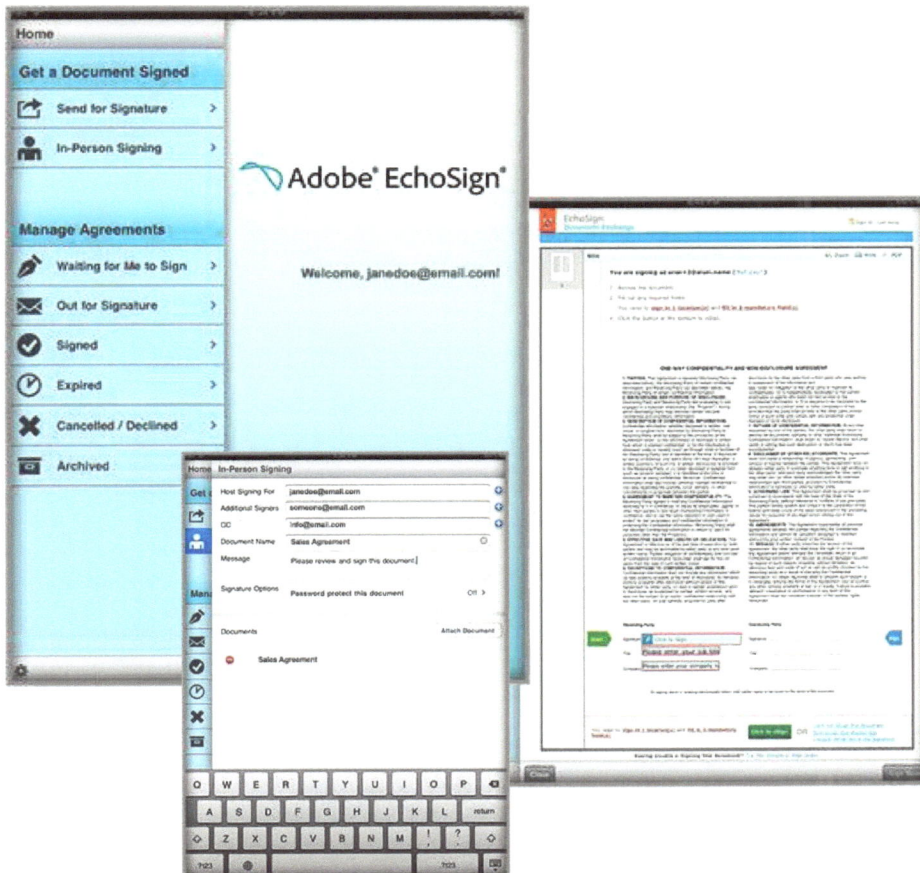

iMeet Mobile

FREE
By **PGi**

Description
iMeet® for iPhone

iMeet is secure video conferencing for business. Brought to you by the Meetings Experts at PGi.

Now from your iPhone, iPod Touch or iPad, you can have an iMeet meeting while on-the-go. Up to 15 people at a time.

With iMeet, there is no expensive hardware to buy, no frustrating software to download and no special bandwidth considerations. Just simple, engaging group video to make all your meetings more enjoyable and productive.

Still teleconferencing? Find out why Inc. Magazine called iMeet "A Better Way to Meet."

PGi

Website:
www.pgi.com
Phone:
866-755-4878
Email:
publicrelations@pgi.com

Company Profile:
PGi has been a global leader in technology, innovation and service for more than 20 years. And it all started in 1991 when PGi founder and CEO, Boland Jones, saw a very simple opportunity: military personnel around the world needed a way to easily connect with loved ones. Jones, along with three others, developed a calling card that was a simple and inexpensive way to connect with anyone, anywhere.

PGi quickly moved beyond calling cards to become a tool for corporate America — constantly innovating new technologies to help professionals do business outside the office. In the last five years, PGi has connected nearly one billion people from 137 countries in over 200 million virtual meetings. But even as the company has grown — it now has more than 1,800 employees — its core has remained the same: to make simple and productive solutions that bring people together.

Awards:
Since its launch in 2010, iMeet has won several industry awards, including:
- 2013 CES Mobile Apps Showdown "Top 10" placement
- Silver "Best New Product" Edison Award 2012
- Nominated for Small Business Influencer Award 2012
- People's Choice Stevie Award 2011
- Best New Product by Best in Biz Awards 2011

"The Making Of" Story:
Meetings are notoriously known as "time and productivity" killers. And, when those meetings include remote participants who can't see who is talking or understand who is on the line, the level of productivity is even more drastically reduced. It's clear that the majority of people would encourage and participate in more virtual meetings if only the tools were easier to use, allowing for better collaboration and improved productivity. This common business phenomenon provided the genesis of iMeet.

iMeet, created by the meetings experts at PGi, integrates best-in-class video and web conferencing with reliable conference calling in a secure, cloud-based, personalized online meeting space where meeting hosts and their guests can get together.

As meetings don't always happen in an office or in front of a computer, PGi developed an iPhone and an iPad apps that allow users to have highly productive meetings on-the-go with iMeet. Just like iMeet's desktop version, users can see everyone in the meeting , chat, take meeting notes and present files right from their mobile device.

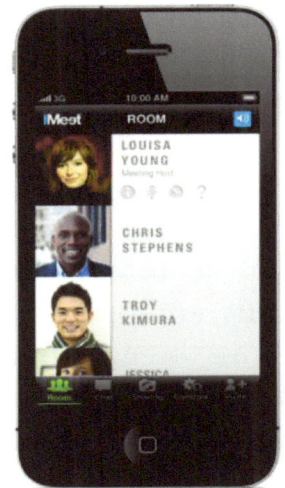

Successes:

Whether connecting from the desktop or the mobile apps, iMeet customers are enjoying the benefits of its simple, personal and mobile virtual meetings.

To date, iMeet HD for the iPad has had 11,300+ downloads and iMeet Mobile for the iPhone has had 15,000+ downloads.

Tips and Secrets:

iMeet was specifically developed to give users a superior virtual meeting experience. iMeet's development team utilized expert insights from social anthropologists to determine how people want to collaborate and drew inspiration from that market research to create iMeet's meeting technology.

Marketing Techniques:

It is important to focus your messages on where your customers do business; and for PGi, it's everywhere. For example, one of iMeet's core user groups is sales professionals. With salespeople spending much of their time on the road, PGi's marketing techniques are focused on showing them how iMeet can be used at any time, from anywhere and across all of their devices.

Risks and Challenges:

With mobile apps, the biggest challenge is deciding whether to build a native app for each platform (iOS, Android, Windows Mobile) or to build a cloud based app using HTML5. There are benefits and detractors to both approaches, so for iMeet, PGi decided to take a hybrid approach. Wherever possible, the team coded using HTML5, but there are certain functional calls that are simply handled better when coded natively.

Setbacks:

Testing is always a challenge when rolling out a mobile application across many different platforms. With the iMeet app, PGi's biggest hurdle involved tunneling through firewalls to support native SIP protocol.

SCOPIA Mobile

FREE
By Radvision, an
Avaya Company

Description

RADVISION's SCOPIA Mobile v3 allows mobile users to connect with full video, audio and H.239 data collaboration to the nearly 2 million installed standards-based video conferencing and telepresence systems worldwide.

Effective Video Conferencing from Anywhere Life Takes You

Users can join standards-based video conferences with full two-way video and see up to 28 participants simultaneously. SCOPIA Mobile can connect to telepresence systems, standards-based HD video conferencing systems and unified communications applications such as Microsoft Lync. Additional highlights include:

> **Radvision, an Avaya Company**
>
> **Website:**
> *www.radvision.com*
> *www.avaya.com*
>
> **Phone:**
> *1.603.666.6230*

- Data collaboration with review capabilities – Users have the ability to view presentations, spreadsheets, documents and images shared in a conference with H.239 data collaboration. Participants can also review previously shared materials without interrupting the presenter using RADVISION's unique advanced data collaboration slider feature.

- Meeting control, moderation and administration – Users can start or stop recording or streaming, lock a conference or end the meeting. Additionally, they can view the participants list and mute background noises, stop cameras or simply disconnect unwanted participants. Users can also change video layouts including rearranging participants. The application also allows users to view statistics such as codecs in use, resolution, network speed and loss for troubleshooting.

- 3G and Wi-Fi capable – SCOPIA Mobile users can video conference almost anywhere through 3G or Wi-Fi including integrated firewall traversal making it a highly effective tool for business travelers. RADVISION's NetSense bandwidth estimation and adaptation technology ensures high quality on mobile internet connections.

Company Profile:

Radvision, an Avaya company, is a leading provider of video conferencing and telepresence technologies over IP and wireless networks. Radvision teams with its channel and service provider partners to offer end-to-end visual communications that help businesses collaborate more efficiently. Radvision propels the unified communications evolution forward with unique technologies that harness the power of video, voice, and data over any network. Visit www.radvision.com, our blog, and follow us on Facebook, Google+, LinkedIn, Twitter, and YouTube.

"The Making Of" Story:

We offer B2B video conferencing and have been pursuing video over mobile for a number of years. With the advent of the iPhone, Android devices and other smart phones and tablets, we saw an opportunity to develop full-featured video apps for mobile devices. Additionally, with BYOD (Bring Your Own Device) becoming a very real phenomenon at companies around the globe, we wanted to equip our customers with an app that allowed them to participate in video conferences from anywhere, over any network and on virtually any device. Our first mobile app offered video conference moderation and was available on iOS and BlackBerry's OS. In October 2011, we launched the industry's first ever business-class video conferencing application featuring full audio, video, content viewing and meeting moderation. The Android version went live in late Q412.

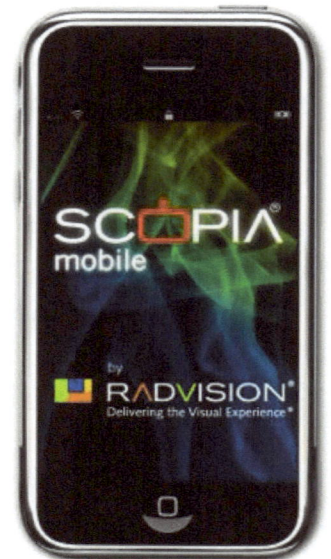

Successes:

We don't make our download info public, but it is in the tens of thousands between iOS and Google Play.

Marketing Techniques:

Scopia Mobile is one of our greatest selling points and often gets us in the door with potential deals. Many companies are talking about mobile video, but most can't make it work so easily, and enterprise-grade mobile video is rarely available without the complexities of licensing and installation. Scopia Mobile just works – click on a link to join a meeting, the app automatically downloads and puts you into your call. You can participate by audio, video and even see live content – as well as moderate a call (e.g. mute participants, add and remove callers, record the meeting, etc.). The biggest tip we have is show people –on-the-spot- how great your app is.

Setbacks:

There were not many setbacks with iOS though we always check for any issues when a new version of iOS comes out. For Android there is huge fragmentation in the market - both in terms of device models and OS versions. As a result, we had to design the application differently for older OS vs. newer OS, and also spent significant time tweaking application for various devices and screen sizes.

PocketCloud Remote Desktop

FREE
By **Dell Wyse**

Description

PocketCloud® is a secure and fast way to remotely connect to your Mac or Windows desktop with your iPad, iPhone, or iPod touch. Access your files, pictures, and applications like Outlook, Word, Photoshop, games or any other program. Simple to install with powerful features and RDP/VNC compatibility, make PocketCloud the best choice for remote desktop access.

Two Simple to Setup Options:

1) Auto Discovery (Recommended): PocketCloud enables easy access to your desktops with minimal setup and no technical know-how. Simply install PocketCloud on your iPad or iPhone and follow the step-by-step instructions.

2) Advanced setup: PocketCloud can also connect directly to machines running RDP or VNC. Simply enter the IP address or hostname to connect.

Dell Wyse

Website:
www.pocketcloud.com
Phone:
408-473-1200
Email:
pocketcloudsales@wyse.com

Company Profile:

PocketCloud was developed and released by Wyse Technology, which was acquired by Dell Inc. in May 2012. Dell Inc. (NASDAQ: DELL), listens to customers and delivers worldwide innovative technology, business solutions and services they trust and value. For more information, visit www.dell.com.

Awards:

- Appy Awards Winner, Productivity Category, 2011, Wyse PocketCloud
  http://www.wyse.com/about/press/release/598
- Mobile Merit Award, Overall Enabler App, 2011, Wyse PocketCloud
  http://www.mobilemeritawards.com/index61.html
- Laptop Magazine, Best of Mobile World Congress 2012, PocketCloud Explore
  http://blog.laptopmag.com/best-of-mobile-world-congress-2012?slide=7
- 148 Apps, Best App Ever, Business, 2011
  http://bestappever.com/awards/2011/winner/busi

"The Making Of" Story:

Wyse Technology has been a long time leader in thin client computing, with a rich set of remote access and virtualization technologies. When the iPhone was announced and SDK made available in early 2008, the company started a project out of the CTO's office to port those technologies to the iPhone. PocketCloud was successfully launched at VMworld in August 2009 and quickly became a top download in the App Store business category. Since then, an Android version, and new product for both Android and iOS called PocketCloud Explore provide a native user interface to remote file access in addition to the view of the remote Windows and Mac desktops.

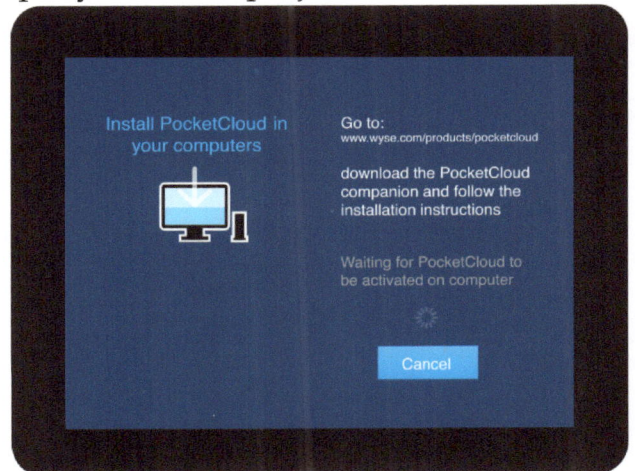

Successes:

We have had nearly 4 million PocketCloud downloads across the App Store and Google Play markets, with a good percentage who paid for the PocketCloud Remote Desktop Pro version at $14.99. It floats among the top 15 Paid and Grossing apps in both the App Store and Google Play Business categories.

Tips and Secrets:

Focus on high quality and earning 4+ star reviews

Best in class feature set, performance, usability, and reliability

Marketing Techniques:

- Maintain an active user forum and support site
- Maintain a strong social media presence
- Run mobile and web page advertising to selected markets, users, regions

Partnership Opportunities:

Open to technology integration or marketing partnerships to address additional capabilities and markets.

Risks and Challenges:

Since Wyse was an established, profitable company, risks were minimized through investment in building the best engineering, QA, and marketing team. Also risks were minimized by taking advantage of Wyse's excellent reputation at product launch.

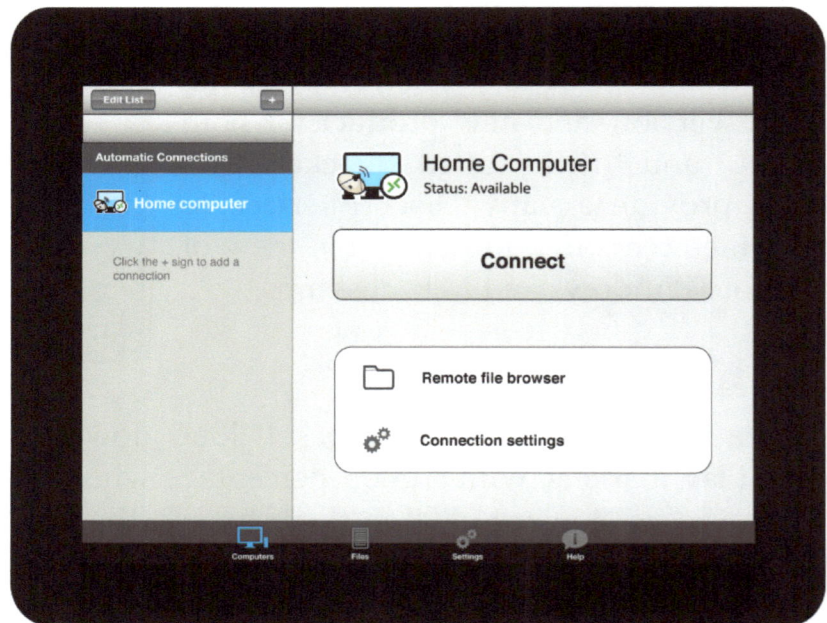

Chapter 9

Productivity

"Procrastinate now…. Don't put it off!"
~ Ellen DeGeneres

MyScript Calculator

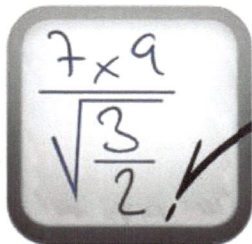

FREE
By **Vision Objects**

Description
MyScript Calculator wins Mobile App Showdown at CES 2013!!

With MyScript© Calculator, perform mathematical operations naturally using your handwriting.

 The Free handwriting calculator for your iPhone or iPad
Easy, simple and intuitive, just write the mathematical expression on the screen then let MyScript technology perform its magic converting symbols and numbers to digital text and delivering the result in real time. The same experience as writing on paper with the advantages of a digital device (Scratch-outs, results in real time, …). Solve mathematical equations by hand without actually having to crunch the numbers yourself.

Vision Objects
Website:
www.visionobjects.com
Email:
contact.marketing@vision objects.com

Company Profile:
Vision Objects is the acknowledged market leader in accurate, high-performance handwriting recognition and digital ink management technology. Vision Objects MyScript™ provides optimum and sustainable results with any digital writing device; it combines digital ink management as well as the accurate recognition of complex mathematical equations and geometric shapes.

With MyScript Smart Note Taking™, Vision Objects turns digital handwriting into interactive data, adding intuitive gesture editing, handwritten notes searches and automatic hypertext linking to external knowledge sources.

Vision Objects solutions are available on all leading desktop and mobile operating systems: Microsoft® Windows®, Apple® Mac® OS, Linux®, iOS® and Android®. Support for 54 languages – including Chinese, Japanese, Korean, Arabic and Hebrew – enables Vision Objects to support worldwide customers from its headquarters in Nantes, France as well as offices in Paris, San Francisco, Tokyo and Hong Kong.

Awards:

Online and Live Winner at the 2013 CES MobileApps Showdown

"The Making Of" Story:

We wanted to apply MyScript digital writing technology to something everybody needs and uses regularly: a calculator. Our goal was to create something different and more powerful than a standard calculator. We realized that the main limit of basic calculators were that using a keyboard it is difficult to edit and make complex calculations. With digital writing, this is typically something we can do: write and edit structured calculations on-the-go. We wanted also to make it as easy-to-use as possible in order to trigger the public's recognition of the power of handwriting, thus giving MyScript a place in the app world and a name in people's mind.

Successes:

The success has been instantaneous as we reached 1,000,000 downloads in one week when we launched the iOS version. Since then, the success is raising day by day with very positive and pleasant (from a developer's point of view) feedback from the users. We receive loads of encouragements and innovative ideas that help us plan for future popular applications. **As of today, we have achieved 3,000,000+ downloads, all stores included.**

Marketing Techniques:

Being reactive in social networks is one thing, another marketing strategy is to launch the app before a big event where you participate (for us it was the International CES) and take part in a competition that will give you good press visibility. Last thing: a good video is always a good communication media support.

Appointment-Plus

FREE
By **Appointment-Plus Online Scheduling Software**

Description

Online scheduling is now mobile! With Appointment-Plus for iPhone, you can take your schedule anywhere with you. Whether your business is mobile or you don't want to be tied down to a computer, Appointment-Plus lets you take your scheduling wherever you work, across the hall, across town or around the world. Manage appointments for yourself or your entire staff, create, update and cancel appointments on the fly. Meet a potential client while you're out, enter their information and set them up with their first appointment from one intuitive interface.

Appointment-Plus is the world leader in online scheduling software, providing solutions for many industries, including health services, personal and professional services, education, events and many others. Appointment-Plus fits any size business from an individual pet groomer to an enterprise with hundreds of locations and thousands of staff.

Appointment-Plus Online Scheduling Software

Website:
www.appointment-plus.com
Phone:
800-988-0061
Email:
info@appointment-plus.com

Company Profile:

Launched in 2001, Appointment-Plus was the industry pioneer in online scheduling software, utilizing the software-as-a-service (SaaS) business model since its inception. Today, Appointment-Plus is the worldwide leader in mobile and online appointment scheduling software with three million appointments booked every month and nearly a billion dollars in commerce conducted annually through its system. Its Scheduling Cloud™ API also enables larger organizations to build custom applications on the powerful Appointment-Plus engine, while its integrated Marketplace allows businesses to easily interface with Microsoft Outlook Calendar, Google Calendar, Constant Contact, QuickBooks and other popular business tools to help businesses and organizations manage their operations more effectively, productively and successfully.

"The Making Of" Story:

As the leader in online scheduling solutions for all-sized organizations, Appointment-Plus is the trend-setter in designing features and functionality that further improves the appointment- and reservation-scheduling processes for its clients. While a few scheduling software competitors offer mobile web apps of their service, none provided true native apps for both Apple and Android devices. As native mobile apps are faster and more easily accessible than mobile web apps, Appointment-Plus determined that native apps would be the most beneficial to clients' scheduling needs. The result is even more efficient appointment and customer management.

Successes:

The Appointment-Plus native mobile scheduling apps have been used by over 9,000 individuals and have booked almost 85,000 appointments through Dec. 31, 2012.

Marketing Techniques:

Appointment-Plus orchestrated both an internal and external marketing push surrounding the release of its native mobile scheduling apps, including: new mobile pages on its Web site; press release; messaging to clients; and pitches to members of the media.

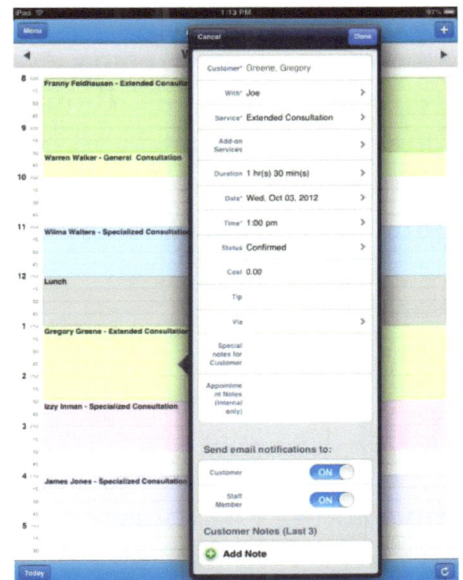

TeamViewer for Meetings

FREE
By **TeamViewer**

Description
Participate in mobile meetings on the go! Join an online meeting from your iPhone / iPod touch or iPad anytime, anywhere. You will never miss an important discussion again: "TeamViewer for meetings" allows you to participate in web-conferences spontaneously and with more flexibility without being in front of your computer.

Download the app, enter the Meeting ID and get started straight away!

Company Profile:
Founded in 2005, TeamViewer is fully focused on the development and distribution of high-end solutions for online communication and collaboration. Available in over 30 languages, TeamViewer is one of the world's most popular providers of online meetings software.

Awards:
TeamViewer received the award "Software of the Year" in 2010 by softwareload.de (leading German download website). The category was "freeware from German vendors"

"The Making Of" Story:
TeamViewer has been originally developed in 2005 as a desktop software for Windows, Mac and Linux. Today the TeamViewer software is being used by more than 100 million users worldwide, and is available in over 30 languages. After having started with a desktop software, we extended our product line over time and

TeamViewer

Website:
www.teamviewer.com

Email:
press@teamviewer.com

developed mobile apps for smartphones and tablets, in order to support our mobile users on the go and to enable them to remotely access their machines whenever, wherever they are, as the number of mobile devices still keeps on rising.

Successes:
TeamViewer is currently being used by more than 100 million users worldwide in over 30 languages. In Google play, we have between 5 and 10 Million downloads.

Marketing Techniques:
TeamViewer has a huge fan community thanks to our Freemium model. Their word-of-mouth helps us to increase our popularity.

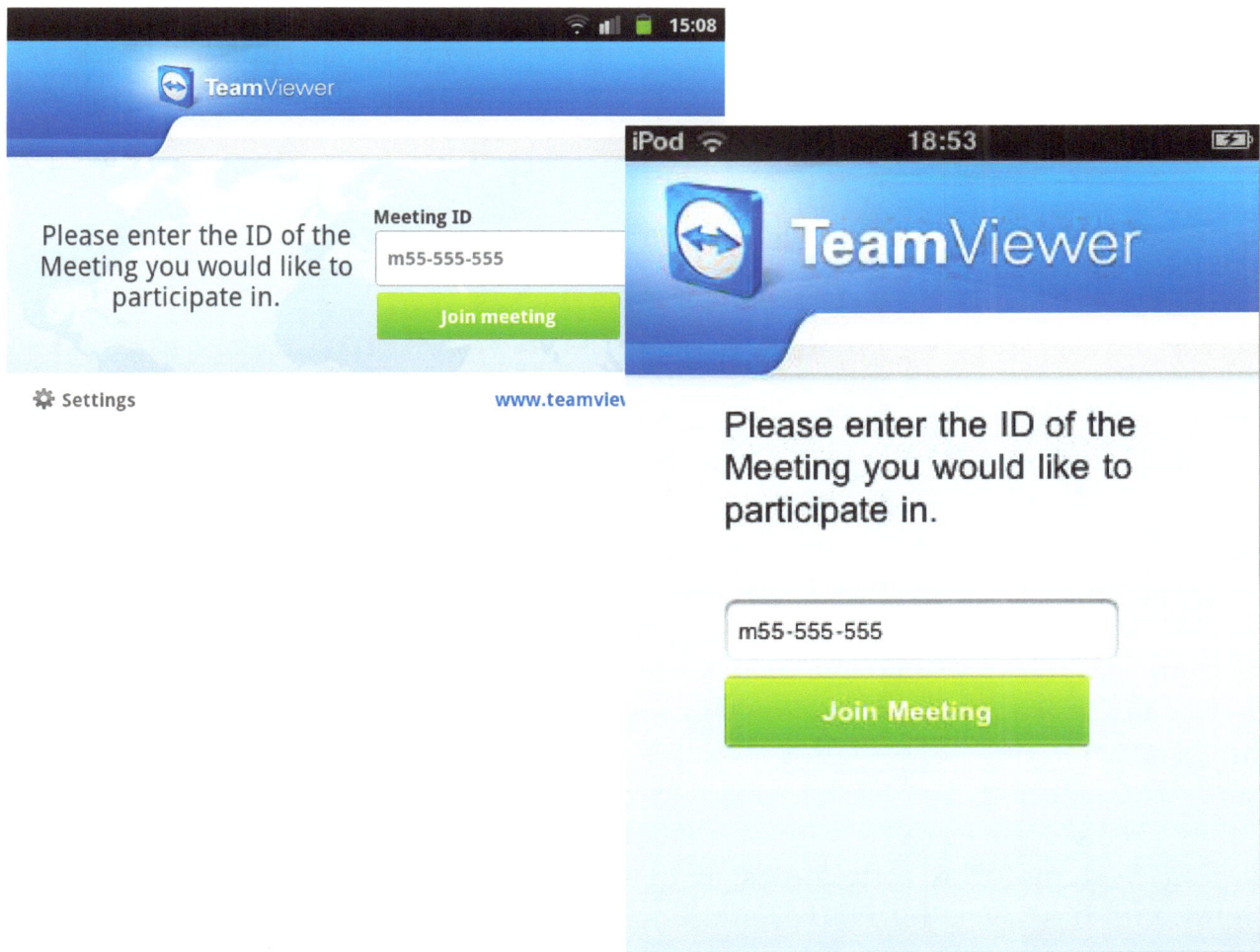

PC Monitor

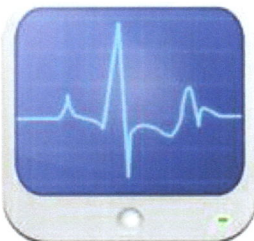

FREE

By **MMSOFT Design Limited**

Description
Monitor and manage your IT infrastructure using this securely encrypted application that gives you total control of your critical servers and applications from anywhere, anytime.

Company Profile:
PC Monitor is an innovative solution for monitoring and managing the IT infrastructure using a securely encrypted mobile app that gives you total control of your critical servers and applications from anywhere, anytime. With apps available for iOS, Android, Windows Phone and Windows 8, PC Monitor is the leading mobile monitoring solution used by over 150,000 IT professionals operating in the SME, Enterprise and managed service and hosting providers sectors. With our easy-to-install software available for Windows, Linux and Mac, our customers are the first to know about any issues affecting their systems and can take immediate action from their mobile device. The original product was developed by a team comprising of leading software developers/programmers and architects led by founder Marius Mihalec. The solution was launched in November 2010 as an iPhone/iPad app and has since been enhanced to support a comprehensive catalogue of devices and Operating Systems. The software is available in Freemium, SMB and Enterprise versions and has over 150,000 user accounts including; Small businesses, Large Enterprises, Hosting and IT Service providers and Managed Service Providers (MSP's).

MMSOFT Design Limited

Website:
www.mobilepcmonitor.com
Email:
sales@mobilepcmonitor.com

"The Making Of" Story:

Up to the creation of our app, there was no other Remote Monitoring and Management (RMM) solution that was truly mobile i.e. supported monitoring devices such as Smartphone's and tablets. In addition, most other RMM solutions were difficult to install and support and required the involvement of highly qualified engineers. With this in mind our development team set about the task of designing a simple, effective and market-leading solution for the "Mobile Age". The PC Monitor app was born in November 2010. The team at MMSOFT has always listened to the most important people in the market, the users, and development has been based on real feedback from our loyal and rapidly growing number of customers. One very important feature designed from the start is the "One License for all monitored systems". Whether your business hosts its servers onsite, in data center's or in the cloud, our app can monitor them all with 1 license. This cuts out the expense and hassle of using multi-site licenses and gives our app a massive advantage over the competition.

Successes:

MMSOFT is a relatively new entrant to the RMM software market (Launched November 2010), however, the number of user accounts now exceeds 150,000 (date Jan 2013). Not only is the company commercially growing at a fast pace but it is also continuing to invest heavily in R&D and 2013 promises to be a massive year for new modules and features.

Tips and Secrets:

"Do your market research and always listen to your customers. When it commercially makes sense to develop a new module or "Plug in", just go ahead and do it".

Marketing Techniques:

Once you have developed an app, get it out into the market and if possible, try to have a freemium version. A freemium version will get your solution commercial traction through early adaptors and provide a perfect market research forum to feed the R&D cycle with ideas for enhancements and additional features.

UX Write - iPad Word Processor

$14.99
By **UX Productivity**

Description

UX Write is a powerful, desktop-class word processor like nothing you've ever seen on the iPad before. Designed for working with large, structured documents, it is ideal for business reports, research papers, technical documentation, theses, and books. While primarily designed for iPad, UX Write also runs on iPhone and iPod touch.

"The Making Of" Story:

Prior to starting work on UX Write, I was working as a lecturer and researcher in the School of Computer Science at Adelaide University, where I had recently completed my PhD. During the process of writing my thesis and research papers, I learned a great deal about professional academic writing, in particular the role that a *structured* approach to document creation plays in both organizing and presenting ideas. I did all my writing using an open source program called LyX, which is based on LaTeX, a structured document formatting system that is very popular in the sciences.

UX Productivity

Website:
www.uxproductivity.com
Email:
peter@uxproductivity.com

When I bought an Android tablet, and looked at the available word processing apps, I was stunned by the lack of functionality that was provided. There was nothing that came anywhere close to LyX or even Microsoft Word in terms of features, and definitely nothing you could write a thesis or research paper in. Not long after, I looked at the iPad and realized the situation there was just as bad. Basically if you wanted to do long-form, structured writing with features like cross-references, numbered headings, styles, and other similar features, you could just forget about using a tablet.

I felt strongly there was a need for a serious, high-end professional authoring tool that academics, students, and business people could use for creating complex documents directly on their iPad without having to resort to a laptop or desktop system.

I began work on UX Write in October 2011, and it took around nine months to get the first release out, followed by another two or three months addressing some initial issues that users had risen. It is now a very stable product, and I've been working on it continuously since its release, adding features like external keyboard support, multilingual autocorrect, and Microsoft Word compatibility.

There's a lot more to come, and I intend to continue developing the app further through the course of 2013 and beyond. I am very active in engaging with users who contact me with suggestions or thoughts on the app, and have incorporated a number of improvements already that have been specifically suggested to me.

Successes:

Since its release in July 2012, I've received a great deal of positive feedback from professional writers who appreciate the unique approach, ease of use, and advanced functionality UX Write offers. Reviews on the app store have, overall, been very good - with many people commenting that it's the best writing app available for the iPad. I know of people who are using it every day for their regular work, and even writing entire books in it.

In its first six months, UX Write has sold around 3,000 copies. I consider this a decent start, and for the past month since marketing has ramped up, it's been in the top 100 grossing productivity apps on the US Store. Once people find out about it and try it out they tend to like it a lot; the challenge is getting the message across to a wider audience, which is starting to happen.

Tips:

- Make use of the outline view to navigate through large documents. This also gives you an overall view of the sections in your document. The outline view is a key feature for working with the structure of your document.
- Use styles to format your document (especially for headings), creating your own custom styles if necessary. UX Write is the only word processor to properly support styles.
- Store all your files on your Dropbox. Not only does this cause everything to be automatically synced between your iPad and computer, but Dropbox also keeps backups of every version you save, so you can go back to earlier versions if necessary.

Marketing Techniques:

At launch, I contacted a number of iPad and other technology sites inviting them to review the app, which some of them did - quite positively in fact. I always offer free promo codes to any journalists or bloggers who are interested in writing a review.

I maintain a blog where I post regular updates, which I use to let people know about upcoming features.

(http://blog.uxproductivity.com)

I've recently begun working with an Internet marketing expert with whom I've put together a new website that drastically simplifies the message of what the app can do, replacing my initial long-winded and overly detailed description of the app. We're using Google Adwords to help attract people to the site, and have been getting a very good response rate from that.

Risks and Challenges:

From a business perspective, I've taken a huge risk in doing this. I've funded the development entirely from my own (modest) personal savings, without any guarantee of success. It's a big challenge to get people to notice your product, and having a good quality app isn't enough - you need good marketing as well. I think the key thing to UX Write's viability is uniqueness - I haven't tried to clone an app, but instead created something new and unique.

From a market perspective, the most significant criticism of the app has been that up until now, it hasn't supported editing of Microsoft Word documents. I chose HTML as the native file format both because it allowed me to use Safari's WebKit layout engine for display, and because it is a well-designed open standard that's supported literally everywhere. But many users will only consider using a word processor if it can work with Word documents; this is perfectly understandable given Word's prevalence in the marketplace, and I've been working on this over the past few months (with a release due in early March that provides Word support).

From a technical perspective, I've had to solve a lot of very complex problems. iOS lacks a user-visible file system, which means I've had to essentially implement my own clone of both the Finder and Dropbox client, neither of which were simple. The editing functionality took a long time to get working reliably. And Microsoft Word's file format is very complex, so it's taken quite a lot of effort to get that working seamlessly as well.

Setbacks:
There have been two major setbacks. The first was that for a long time, the app was selling very poorly. Despite being highly usable and stable, and the fact that I was getting very good feedback from people who were using it, it was just getting lost in the noise. Most of my competitors had much better marketing, SEO, and brand recognition - I've learned the hard way that this is what makes money on the app store, much more so than any technical merits of an app. Fortunately I've found someone now who is good at these things and is working with me to promote the app, and we're starting to see some good results from these efforts.

The second major setback has simply been the amount of time it's taken to implement various features. Programmers *always* underestimate the amount of work involved with a given task, since software development is highly unpredictable and you only find out how long something takes once you've done it. Having said this, I believe the attention to detail, usability, and quality I've spent this time on has paid off in terms of a very high quality app.

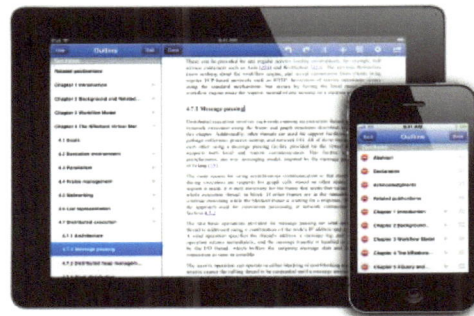

iKnowU Keyboard

$1.99
By WordLogic

Description

iKnowU's award-winning keyboard is the most feature-rich and intuitive you'll ever use. Using state-of-the-art patented technology it knows what you're about to type - letting you complete sentences at blazing speed. It's so accurate, you'll be amazed you ever typed without it!

iKnowU becomes more accurate every time you use it, constantly learning about you and your style. Amazing features such as WordChunking™ and Gesturing™ enable you to chain together phrases and create whole sentences in a matter of seconds, saving you precious time. iKnowU is more than just a keyboard, it's an extension of your personality.

WordLogic

Website:
www.wordlogic.com
Phone:
604-257-3660
Email:
getinfo@wordlogic.com

Company Profile:

WordLogic Corporation (otcqb:WLGC) develops, markets, licenses and sells advanced predictive platform software designed to accelerate information discovery and text input. The Company's innovations operate on a wide variety of devices, including smartphones, PCs, cell phones, Smart TV, media players, automotive navigational systems, infotainment, and game consoles. The Company's intellectual property portfolio includes six issued U.S. and European patents and three pending U.S. patent applications.

"The Making Of" Story:

We've been developing and patenting predictive text technology for over 10 years, so we know there isn't another keyboard app on the market that can match what iKnowU does. Anyone looking for a simple and incredibly effective keyboard app will love iKnowU. In terms of accuracy, speed and ease of use, there's nothing else to touch it. The app intelligently predicts not just words and phrases, but complete sentences as you type - allowing you to add them with a simple gesture. Not only that, but the app can learn as you type, becoming even more accurate the more you use it.

Marketing Techniques:

We employed a specialist app PR and marketing agency, Dimoso, for the launch of the app. They distributed exclusive content to the media in the US for our national launch and will replicate the strategy for our worldwide release.

TeamViewer for Remote Control

FREE
By **TeamViewer**

Description

TeamViewer provides easy, fast and secure remote access to Windows, Mac and Linux systems.

TeamViewer is already used on more than 100,000,000 computers worldwide and with the TeamViewer App for iPhone and iPod Touch you are able to:

- On the fly support your family and friends

- Have access to your own computer with all of its documents and installed applications

TeamViewer

Website:
www.teamviewer.com

Email:
press@teamviewer.com

Features:

- Remotely access unattended computers
- Conveniently control remote computers using the iPhone multi-touch gestures: left click, right click, drag & drop, scroll wheel, zoom, change monitor
- Complete keyboard control incl. special keys such as Windows®, Ctrl+Alt+Del
- Remotely reboot the computer
- Automatically adjust the screen resolution of the remote computer
- Overview of computers that are online via the integrated Partner list
- Effortlessly access computers behind firewalls and proxy servers
- Meets highest security standards: 256 Bit AES Session Encoding, 1024 Bit RSA Key Exchange

Chapter 10

Reference

"Personal beauty is a greater recommendation than any letter of reference." ~Aristotle

ThankYouPro Thank You Cards & Giftcards

FREE
By **ThankYouPro**

Description

ThankYouPro is the business app that takes the effort out of sending relationship enhancing cards and thank you notes. ThankYouPro lets executives, salespeople, brand representatives, product distributors and many other people who interact frequently with customers and partners use their smartphones to easily send fully customized thank you cards via the US mail.

Created by John Testement in 2011, Colorado-based ThankYouPro is an application aimed at allowing sales professionals to create thank-you cards equipped with elegant designs, a handwritten signature, personalized logo or photo and rapid delivery times.

With over 20 award winning, professional designs, or the ability to use photos from your iPhone photo album or camera, ThankYouPro has dedicated itself to helping consumers create elegantly designed thank-you cards for a variety of situations.

ThankYouPro

Website:
www.ThankYouPro.com
Phone:
206-369-3432
Email:
johnt@thankyoupro.com

How does ThankYouPro work?

After downloading the app, ThankYouPro users have the option of picking one of 23 different designs or selecting a picture from their iPhone photo album and positioning it as they please. Simple customization steps let people modify their personalized messages choosing from a wide selection of handwriting fonts, sizes, colors and text alignments. Advanced 'electronic ink' technology lets you write your own signature right on the face of the iPhone screen to be placed on the card. The

ThankYouPro printing process prints the messages on cards in a tactile manner that resembles actual handwriting and ink. Cards can be further customized with company logos that are likewise sized and positioned to the sender's liking. After the card is formatted, the recipient is either selected from the phone's address book or entered manually on the spot, and with a tap of a button the card is mailed. A first class stamp is added for fast professional delivery – no bulk mail permits. ThankYouPro also offers an option to send free cards via email.

Company Profile:
ThankYouPro is a small privately owned business. John Testement founded ThankYouPro after teaching himself iPhone programming as a hobby. Having started another web-based business that used on-demand printing for Thinking-Of-You cards, John and his wife Leslie came up with the idea of combining mobile technology with on-demand printing to send Thank You Cards. To avoid competing in the general greeting card business John decided to create a more professional product suited for sales professionals in real estate, financial service and other vertical markets.

John Testement is an entrepreneur, engineer, marketer, consultant, and coach. John has over 20 years of marketing management experience in high technology organizations such as Texas Instruments, Philips Corporation, and Aldus/Adobe Corporation. His work in large firms is complemented by years of experience running his own businesses. John founded Zendoa Corporation in 2011 to market ThankYouPro.

"The Making Of" Story:
John liked building things. While running an executive coaching and consulting business for the last 18 years, John also built 2 airplanes and a home theater. Needing another project after the theater, John decided to learn iPhone programming for fun. Then the idea for ThankYouPro came along and John realized the potential of giving busy sales professionals a convenient tool for doing what they know makes a huge impact on customers - sending thank you cards. ThankYouPro could allow a sales person to create a thank you note, before leaving the customer's parking lot, to be in the mail the next day

Successes:

We have won some great corporate accounts. App sales continue to grow despite a fairly small investment in marketing.

Tips and Secrets:

We make sure every customer is delighted with their experience using our app and our service. So far every customer who has emailed or called us with a problem has ended giving us a 5 star review!

Marketing Techniques:

Most of our new customers come from Apple App Store searches and word-of-mouth. We also do a small amount of Google and Flurry advertising. We have tried using PR firms with disappointing results.

Risks and challenges:

The biggest challenge is getting visibility when there are over a million iPhone apps. Fortunately, we have strong word-of-mouth marketing. Our customers are spreading the word which is creating our growth.

Setbacks:

We did spend a lot on PR firms that produced disappointing results. That used up a good portion of our launch budget. We found that some PR firms over promise and under deliver.

At first we thought that Apple's Cards app would be a significant competitor. As it turns out they helped to validate the iPhone-produced printed card market. Apple's product and service received a high percentage of poor reviews which gave us great insight into what customers really wanted. By doing all the things right that Apple did wrong, we now enjoy 99% 5 star reviews!

Choose Card   Write Note   Add Address   Mail It!

Chapter 11

Utilities

"Utility is when you have one telephone, luxury is when you have two, opulence is when you have three - and paradise is when you have none."
~Doug Larson

Norton Mobile Security

Powerful Protection for your Mobile Devices

$29.99/year
Norton by Symantec

Available on the App Store

ANDROID APP ON Google play

Description

Norton™ Mobile Security is a web-based service that helps you remotely lock and locate your lost or stolen Android mobile device, iPhone and iPad. You can also back up your contacts and easily restore them across your mobile devices. To use this app you must register for a Norton Mobile Security subscription. Start protecting your devices today.

| Back | Backup | Restore |

Contacts in Cloud — 8

Contacts on Device — 8

Last backup — 09/24/2012 10:43

Backup

Norton by Symantec

Website:
http://us.norton.com/norton-mobile-security/

FEATURES

- Find your lost or stolen iPhone, iPad, or Android device on a map with remote locate (requires an active 3G or 4G data plan)
- Back up your contacts and easily restore them or share them across all your mobile devices
- Control the security for all your iOS and Android mobile devices with easy-to-use and convenient web-based management
- Protect Android devices from mobile threats including viruses and malware

My Clock Station Pro

FREE
By Game Scorpion Inc.

Description

My Clock Station PRO is your all in one productive clock solution! This app is designed to be among the most functional clock apps in the market!

With so many features, it's sure to please!

FEATURES:

- Beautiful HD Clock with BIG BOLD Numbers!
- Built-in Calculator
- Full Sleep Therapy solution with 10 Built In sounds including Ocean Waves and even Crickets!
- Notes/Memo section
- Flashlight (Turns screen white so you can use device like a light)
- Screensaver (To help avoid screen burn)
- Various Themes to Choose From
- Nap Timer
- Dual Alarm

Game Scorpion

Website:
www.gamescorpion.com

Email:
info@gamescorpion.com

TouchPal Keyboard

FREE
By **CooTek**

Description
TouchPal Keyboard, Feel The Speed!

Annoyed by typing without accurate prediction for multi-languages? Eager to try the new sliding input instead of tapping?

Now you have a choice! TouchPal brings brand-new keyboard to iPhone, with the best contextual prediction and sliding input (TouchPal Curve®).

TouchPal:

- The most innovative and the best keyboard.
- Own The Global Champion of GSMA Mobile Innovation Award.

TouchPal Keyboard helps you to type faster when you reply to short messages and write emails.

CooTek

Website:
www.touchpal.com/en/index.html

Email: *karl.zhang@cootek.cn*

Company Profile:
Founded in 2008, CooTek is dedicated to improving the usability, functionality and overall value of mobile devices via the innovative software we design and bring to market. Its TouchPal Keyboard application for Android and iOS is based on CooTek's multiple patented technologies and delivers a powerful, easier-to-use alternative to traditional hardware keyboards.

Awards:

CooTek was awarded the GSMA Mobile Innovation Champion in 2009 and was most recently recognized in 2012 as a final-stage company in TechCrunch Disrupt's Battlefield in Beijing. In addition, TouchPal Keyboard has been featured twice by Google on the Google Play main page as the most innovative keyboard for Android.

"The Making Of" Story:

As touchscreens are standard for most smartphones and tablets today, many consumers are familiar with the "fat finger" phenomenon. Pre-installed keyboards can be sensitive and equipped with inaccurate automatic correction tools. As such, we wanted to create a keyboard that enables users to feel more comfortable with touchscreen technology with a keyboard that could adapt to them, instead of vice versa. TouchPal Keyboard is the only free, intelligent keyboard that uses enhanced swiping technology so users can type without lifting a finger. It can also detect your personalized style of typing through its blind typing and learned predictions features, ensuring a "fat finger" is never an issue.

Successes:

Before launching as an Android and iOS app, TouchPal Keyboard became a default keyboard on the HTC One X and New Desire devices, the Sony Xperia Ion, Huawei Honor and Ascend devices, and ZTE Fury and Blade devices. After its release on Google Play and the App Store, TouchPal Keyboard was also featured in leading publications like Lifehacker and The Next Web. CooTek has received outstanding press coverage proving TouchPal's popularity and success over competitors; since its launch in November, it has been featured in over ten top mobile and technology publications and has reached over 3,000,000 downloads on Google Play and 200,000 in the Apple App Store.

Tips and Secrets:

Our biggest tip is to understand user's pain points and design an innovative solution solving those needs. CooTek has been focusing on the input technology on touch screens since 2008 and observes the pain points of the users carefully. As such, every innovative feature is designed to solve one of those pain points.

Marketing Techniques:

One of the biggest challenges when developing an app like this is recognition in the market. In order to combat that, we focused our marketing efforts into comparing our keyboard directly with competitors to prove that TouchPal is a superior keyboard. For example, one feature in Talk Android dubs TouchPal as an app with the combined advantages of SwiftKey X and 8Pen along with the predictive typing missing from the Swype keyboard.

Risks and Challenges:

One of the most prominent challenges when launching TouchPal was the fact that other intelligent keyboards already come pre-installed on Android devices, like Swype. There are other well-established intelligent keyboards available for download as well.

Setbacks:

One of our biggest hurdles was the fact that TouchPal Keyboard cannot be used as the default keyboard on iOS devices. Apple does not allow the installed keyboard to be replaced by third-parties, so we needed to create a standalone keyboard app that would be appealing to users. We achieved this through one-touch export buttons to Twitter, Email and most notably, Evernote. Now, during a meeting or in school, users can swipe their notes and immediately export it to their Evernote account, making note taking easier and faster.

ShopAdvisor

FREE
By **ShopAdvisor**

Description

ShopAdvisor keeps track of products you are interested in, and notifies you when the deal is right.

ShopAdvisor does this for more than 8 million users by crawling through more than 16,000 online and local retailers. ShopAdvisor has already sent price alerts worth more than $38 million in savings on more than 700,000 products.

- Never miss a deal - when the price drops, ShopAdvisor notifies you.
- Know if the price is right – product information includes a 6-12 month price history
- Make informed decisions – ShopAdvisor has expert advice on all products
- Anywhere, anytime access – add to your list, check your list, and get alerts directly on your iPhone and on your desktop web browser.
- Local data - also see nearby deals from brick-and-mortar stores in your neighborhood

ShopAdvisor

Website:
www.shopadvisor.com
Phone:
617-986-5022
Email:
contact@shopadvisor.com

Profile:

Scott Cooper, CEO of ShopAdvisor, has grown a diverse set of businesses from inception to $100M+ in revenue, grown large business segments through acquisition, and built relationships with companies of all sizes. During his career he has led multiproduct business units, large global development organizations, and one of the world's largest indirect channel marketing organizations. He is a pragmatic leader with a passion for identifying world-class talent and innovative business ideas that lead to value creation for shareholders.

"The Making Of" Story:
There's a lot of distance between interest in a product and actually buying it and consumers almost always wait before making a purchase. They may see a product in a magazine, on a website, or in a store and like it but most do not take out their wallet and buy the item right away. Instead, they wait until they have enough cash, until they can ask a friend, or until the price is right.

ShopAdvisor is the only platform designed around this natural shopping behavior. Consumers can "watch" products they've discovered and later, when the price is right, ShopAdvisor reminds them of that product- on their phone, in their email or on their tablet.

Successes:
ShopAdvisor has found more than $38 million in savings on more than 700,000 products for its more than 8 million active users.

Tips and Secrets:
So many mobile apps suffer from the one-and-done reality of consumer usage and all too often downloads rarely become active users. Because ShopAdvisor is built around the notion of alerting users later of prices that shift on products, there is a built-in mechanism to drive continued usage. The key to the success of this approach is that all alerts sent to users include specific value (cost savings) that the users themselves have requested. ShopAdvisor's success hinges on meeting this natural human behavior about how people shop.

Marketing Techniques:
The ShopAdvisor product "markets" itself to its users. Every time a user "watches" a product in ShopAdvisor, the App will re-engage the user in the near future. This has led to an average 4.6 items "watched" per user, and gives ShopAdvisor a very high active user ratio.

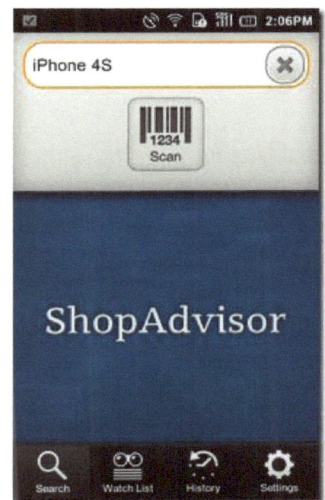

Partnership Opportunities:

Yes. ShopAdvisor works with multiple third-parties to integrate our shopping service within their smartphone, tablet and web apps.

Risks and Challenges:

Like all developers of mobile apps, getting visibility among the ocean of mobile apps in an app store is highly challenging. Although we have had great success having been featured prominently in several app stores, it has always been clear that we need additional ways to acquire new users. After the completion of our iPhone, iPad and Android versions, we decoupled the basic service from the app itself and embedded ShopAdvisor within independent websites/blogs, and within the tablet editions of magazines. This allows us to leverage the appeal of well-placed curators of highly targeted audiences.

Setbacks:

We entered a "partner app" competition with the intention of earning good visibility with a popular brand. This required us to shift development priorities even before we officially launched our initial iPhone app. Although we did not win the competition, which was a setback, the development work was essential to our ability to later decouple our shopping service to work within independent websites and tablet editions of magazines. In typical fashion, the setback sowed the seeds of future opportunity and success.

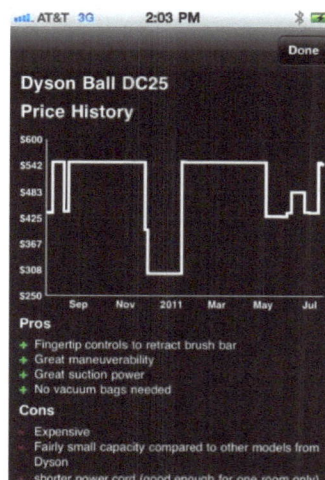

RingMindMe

FREE
By **Peggywrites, LLC**

Description

You turn off your iPhone ringer for a meeting or a movie (that's a good thing). Later, you miss calls because you forgot to turn your ringer back on (that's a problem). RingMindMe is the solution.

RingMindMe is a "set it and forget it" ringer manager. You set a time-of-day schedule or choose your current location, turn off the ringer and RingMindMe alerts you when it's time to turn the ringer back on.

With RingMindMe, you won't be embarrassed during a concert, class, or meeting and you won't be frustrated by missed calls after the event is over, or after you leave the event's location.

"The Making Of" Story:

I was frustrated that after I turned off my ringer for a meeting or other event, I missed calls later, because I forgot to turn it back on. Then, I talked to friends and co-workers, who all told me they had the same issue. I looked for an app, and when I didn't find one, I decided to make one so life would be better for all of us (or at least, all of us with iPhones).

The app is easy to use, either by time or location. When you choose the Time option, you can use either the default options (Ringer off Now, Ringer on in 1 hour) or the familiar "click wheel" interface to set specific off and on times. When it's time to turn your ringer on again, your phone will vibrate, and display a message on your screen.

Peggywrites

Website:
www.ringmindmeapp.com
Phone:
650-269-1883
Email:
Peggywrites@yahoo.com

If you need more "Ringer off" time, just tap Extend to delay the message for 15 minutes. Like a "snooze" button, you can use the Extend feature as often as necessary.

To use the location option, just tap Start, set the size of the zone where you want your ringer off, and as soon as you leave the zone, RingMindMe alerts you to turn your ringer back on. The zone can be as small as a building, or as large as a neighborhood. Either way, once you get a short distance from the zone, you will get an alert. Of course, you can expand the zone, if you like, when you receive the alert.

Don't worry that you will miss the alert. Whether you used the Time or Location option, RingMindMe will alert you once a minute until you respond.

Successes:
My app received a great review from Jim Dalrymple's The Loop website and has been featured on websites such as Shoestring Venture. When my clients see me use the app, they not only want to get it, they are even more impressed that I have my own app.

Tips and Secrets:
Get your friends to beta-test any new app. They will give you really valuable feedback that will make your app much better!

Marketing Techniques:
Word of mouth is still the most common way people buy apps, so getting a buzz going is very important. Talk about your app to friends, strangers, people you meet, anyone who will listen. And respond to relevant HARO queries to get media attention for your app.

Setbacks:
At first my app was rejected by the Apple App Store, so I made some changes, including adding the location feature. When my beta-testers tried out the app, they were not able to figure out how to use some of the features, so I changed some of the graphics and the layout of the app to make it more intuitive. Those changes worked, and later beta testers really liked the app.

ShopSavvy (Barcode Scanner and QR Code Reader)

FREE
By **ShopSavvy**

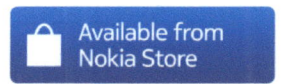

Description:

Find the right product at the best price! Now the fastest, most accurate and most comprehensive scanner around. It won't find everything in the world, but it is close. Now you can be savvy when you shop!

Company Profile:

ShopSavvy is the world's leading mobile shopping platform with more than 30 million downloads and millions of active users. Available on the iPhone, Android and Windows smartphones, ShopSavvy enables consumers to scan the UPC, EAN and QR barcodes of products they want to buy and do immediate, in-store price comparisons with local and online retailers. Users also can find product reviews and other relevant information about scanned products.

Awards:

ShopSavvy has won numerous awards and recognitions, including the Google Android Developer Challenge, Netexplorateur of the Year, Mobile Monday Mobile Peer Awards finalist, Under the Radar's Best App, Microsoft BizSpark Accelerator, TechCrunch Crunchies, CES Top 20 Mobile App, Best Shopping App Award from APPScar, Tech Titans Emerging Company Horizon Award finalist and Online Retail Awards finalist.

ShopSavvy

Website:
www.shopsavvy.com
Phone:
972.752.3832
Email:
rylan@shopsavvy.com

"The Making Of" Story:

ShopSavvy began in 2008 when a company co-founder learned about a mobile shopping application called GoCart. It was to be among the top winners in the first Google Android Developer Challenge. GoCart was renamed ShopSavvy, and new platforms, versions and features followed, culminating with ShopSavvy.

Successes:

ShopSavvy has been downloaded more than 30 million times and has millions of active users. ShopSavvy's latest goal is nothing less than a revolution in retailing. Fueled by a $7 million round of funding from investors including Facebook co-founder Eduardo Saverin, ShopSavvy aims to transform itself from a shopping application to an all-encompassing shopping community.

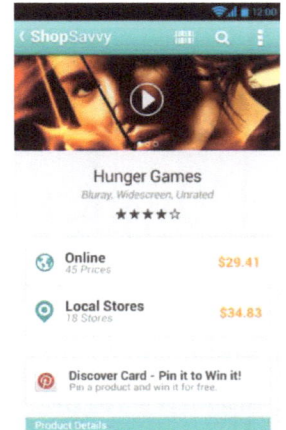

Tips and Secrets:

There's no secret really. We're focusing on a difficult problem and on solving it very well to the exclusion of all else.

Marketing Techniques:

ShopSavvy has been featured thousands of times by popular media outlets around the world and the number of media mentions continues to increase. Company co-founders Rylan Barnes and John Boyd often travel to mobile- and retail-related conferences to speak about their experiences building ShopSavvy and how retailers can reach audiences in more targeted ways through mobile advertising.

Risks and Challenges:

We've taken risks moving into adjacent-use cases beyond just where to buy something; and we'll be excited to see which of those adjacencies have the greatest impact on the shopping experience.

Setbacks:

As the market has matured, we've come to believe a direct relationship with our customers has become more important. In the beginning, we didn't even ask for their email addresses. As we've developed into more of a shopping platform, it's been a trade-off between asking for more information from that user and what they've expected from us in the past. This transition, while probably characterized as smooth now, had its rocky points as consumers changed how they viewed our application.

PART 4: Education and Reading

Chapter 12

Books

"I just wrote a book, but don't go out and buy it yet, because I don't think it's finished yet."
~Lawrence Welk

PlayTales Gold Bookstore

Where kids read & play with interactive children's books

FREE
By **PlayTales**

Description

Unlimited access to entire PlayTales Catalog, featuring classic children's tales, educational material, and more!

- Read your favorite stories even without an Internet connection.
- Different reading modes to choose from: Read to me, Read by myself, or Autoplay.
- Original music, sound effects, interactive features.
- New stories every Friday!
- Activities, Coloring and Songs.
- Universal subscription between iPhone and iPad devices.
- Stories available in 8 languages (English, Spanish, French, German, Italian, Portuguese, Chinese and Japanese).

PlayTales

Website:
www.playtalesbooks.com
Phone:
617- 322-9863
Email:
info@playtalesbooks.com

Get all the stories you want with original music, sound effects, interactive features, and more!

Company Profile:

PlayTales produces award winning apps featuring unlimited interactive books for kids. PlayTales' interactive books are designed to inspire, educate, and entertain readers ages 1 to 8 and stories are available in eight different languages. PlayTales was founded in 2010 as a wholly-owned subsidiary of Genera Interactive, an internationally based multi-platform mobile entertainment and utilities provider with offices in the USA, UK, Spain, Romania, and China. PlayTales apps can be downloaded through iTunes App Store, Google Play, Amazon.com, Blackberry App World, Intel App Up, and Verizon V-Cast.

PlayTales offers users the highest-quality interactive children's books currently available for mobile devices with in-app games and activities that help stimulate children's minds. Our apps feature classic children's tales such as The Ugly Duckling, Sleeping Beauty, and Puss in Boots, along with new favorites like Pocoyo and Felix the Cat. PlayTales' stories also feature educational content that help children learn how to count, read, spell, add, subtract and more.

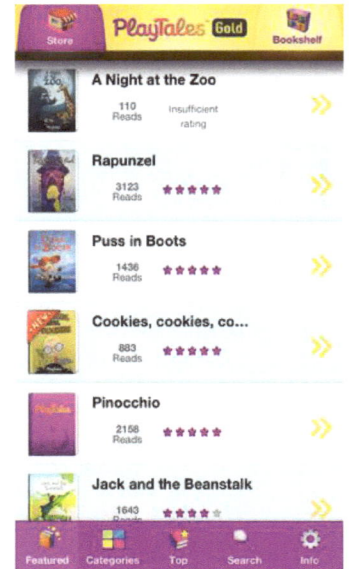

In 2012 PlayTales became the world leader of interactive children's books with over five million monthly reads and three million users worldwide. In the last two years many companies have worked with PlayTales' creative teams in collaborative projects to produce top-quality content for users. Some of PlayTales' major partners include: Harper Collins, Hachette Book Group, UNICEF, and Fleischer Studios, Inc.

Making of Story:
PlayTales was created to fill a void in the mobile app world; as more and more eBooks were being developed and sold, we noticed that there was a lack of interactive books for children that could be accessed through mobile devices. Why should parents have to spend more money buying an eBook reader specifically for their child when they could just hand over their Smartphone or tablet? We wanted to give families the opportunity to read not only entertaining books, but educational ones as well, anywhere and anytime they wanted. We also added interactive features to our stories to hold children's attention and make them more interesting to young readers. This is why PlayTales was started; to provide families with a wholesome kid-friendly application that provides unlimited, top-quality stories that can be accessed at anytime and anywhere in the world.

Successes:

In October 2012 PlayTales achieved over 5 million monthly reads; this means that through our PlayTales apps our interactive books were read over 5 million times by users. We also recently accumulated over 3 million users worldwide and the number continues to grow daily.

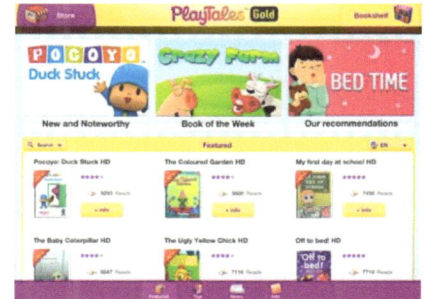

Tips & Secrets:

We wanted to give our users the highest-quality interactive books available—so we developed our own digital publishing technology to make that happen. We are constantly developing and improving our publishing tools to give users the best reading experience possible. The quality alone of our books is what separates PlayTales from the competition—when you compare us to other bookstore apps the difference in image resolution and interactive features is clearly visible. Also, aside from content quality, localization is a very important part of the success of PlayTales too. Our apps are available worldwide and we translate stories into eight different languages—another unique feature which sets us apart from our competition in the app market. We always make sure to use local translators who help guide us in the story design and wording for each lingual version of a story. Localization is probably one of the most difficult parts of being an international app because it can make or break a story; if a book isn't correctly translated or culturally relevant to the audience, the book will not succeed.

Marketing Techniques:

Our marketing efforts are focused on communicating with parents and users. We contact Mommy Bloggers, tech media, and various app sites by submitting our apps for review to gain brand recognition through positive reviews and word of mouth. We also regularly maintain our blog and social media sites to promote consumer interaction, relationship management, and raise brand awareness. We've found our social media sites to be very useful when dealing with consumer questions and issues.

Risks & Challenges:

Before PlayTales there was TouchyBooks…originally PlayTales was named TouchyBooks when it was founded in 2010. Once we became big enough for international distribution, our partners and advisors recommended we change our name—the word "touchy" had a negative connotation in the USA and if we wanted to be successful worldwide it had to change. In February 2012 TouchyBooks officially became PlayTales. This was a huge risk for us because all of our users were accustomed to our social media sites, website, and apps featuring our former name. Also all of the press coverage, reviews, and media mentions we had accumulated featured the name TouchyBooks…we risked losing all the recognition and status we had worked so hard to obtain by changing our name. Through serious communication efforts we have been able to convey to the public that even though our name has changed, we are still the same developer and they can expect the same top-quality stories they have become accustomed to.

Chapter 13

Education

"I hope you're pleased with yourselves. We could all have been killed – or worse, expelled. Now if you don't mind, I'm going to bed."
~ Harry Potter and the Sorcerer's Stone

Astroloquiz

FREE
By Together Learning Media, Inc.

Description

Astroloquiz provides a fun learning experience based around the 12 signs of the zodiac. Over 90 definitions of personalities are presented in a way that young learners and old alike can easily understand the meaning of personality adjectives.

Young learners will be intrigued by how their friends will react to the 48 separate scenarios presented in this book. It is a great springboard for discussion about how we all react differently to the situations we face every day. Having infinite replay value, the quiz itself takes about 30 minutes to complete.

Take the Astroloquiz today!

Company Profile:

Together Learning Media Inc. was founded in 2012 by Dave Wingler, a teacher whose first applications created in partnership with INKids Pty. Ltd., Futaba Word Games and Futaba Classroom Games which were met with great acceptance by not only students and educators but have also been featured by Apple three times in the past year. Operating with the mission to create learning applications that provide educational and social learning opportunities for use in K-12 classrooms, we provide opportunities for educational professionals and their institutions to bring collaborative play into the technological integrated classroom.

Together Learning Media

Website:
www.togetherlearningmedia.com
Phone:
843-628-2951
Email:
dave@togetherlearningmedia.com

Astroloquiz
What's your REAL star sign?

- Narrated by professional voice actors and actresses.
- Learn over 90 personality adjectives.
- Stimulates social awareness and engaging discussions.

One of the most difficult challenges a teacher faces is trying to engage the whole class at one time. No matter the class size, the fact that every student is different makes it difficult to find the perfect way to pitch ideas or topics that will make everyone in the class interested. The technology laden classroom may find students even more isolated in their own device during class time. There are plenty of lesson plans and topics that enable students to legitimately and beneficially work by themselves, there is definitely a need to bring students together in an interactive fashion. Getting youngsters working together adds a social dimension to their learning and it also provides students with the chance to learn about each other.

The use of apps in the classroom has really helped teachers in a number of ways to provide a better classroom experience. Finding apps that provide educational benefits alongside interactive elements really helps a teacher to create something which provides a much valued learning experience and is also engaging for all of the students.

Making of Story:
I'm a teacher. Astroloquiz started out in my classroom as a pencil and paper way for students to learn more about themselves and their peers while at the same time learning language. Thanks to the iPad I could bring it alive! Astroloquiz uses the personality traits common to each star sign in the Zodiac as a springboard for discussion. The app itself provides 48 unique scenarios, four for every star sign, and is aimed at encouraging young learners to explore and understand their personality and the personality of those around them all the while developing their conversational ability. Over 90 adjectives are defined and have audio recordings for them in a

definitions section in the app. To promote listening skills, 8 separate voice-over artists who have recorded with big names like Disney and NASA, were involved in the project, bringing to life every character.

In an educational environment for young learners, the app is intended for children in the 5th or 6th grade. My hope is that it can be used by everyone as the quiz is fun and engaging. To be honest with you, as a teacher I've been able to create something that I feel others may find delight in using. My goal is to help others learn without them actually 'thinking' they are learning. It's an indirect approach and has 2 outcomes. People learn meanings of words while at the same time learn about themselves and others. It is this spirit of personal discovery that I like to create.

There is a lot to be said for producing bright youngsters who can make decisions and think about the situations that face them. Educational bodies and schools are placing a greater emphasis on this form of development and it is good to lead the way with respect to helping children develop their personal outlook on life. Promoting collaborative work between students will provide a great start in molding the adults of tomorrow.

I think any parent that believes a school is not doing enough to prepare their children for the situations they will face in the "real world" will be delighted to know that there are educational apps providing them with the opportunity to explore more than core learning plans. In this sense, the iPad or any other tablet is the ideal accompaniment to the learning process, ensuring that every student can receive a well-rounded education which touches on every aspect of life.

The Astroloquiz is not intended to be the most important educational app in a student's life, but it is perfectly aimed at complementing the more traditional lessons that are contained in educational apps.

Marketing Techniques:
I attempted to have a coordinated launch for this app based on what I had read from other developer's experiences. I've used Twitter, Facebook and Pinterest in marketing the app. While many people have expressed support through each of these social media portholes, this has not translated into sales. I also budgeted and coordinated for Astroloquiz to appear on a number of App Review sites in prominent

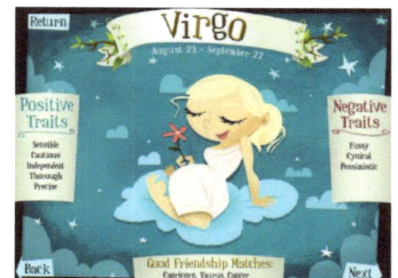

advertising positions. The move was actually more costly than some aspects of developing the application. This strategy also has had absolutely no effect on sales. Additionally, while reviews I have received for this application- both paid and non-paid have been positive- this too has had little influence on sales.

Risks and Challenges:

I started my business for posterity, for the sake of my students and future students. I wanted to create a learning activity that could be used countless times over many generations to enable language learners to learn more about themselves while studying language. I didn't have the implicit expectation that I would obtain an ROI but decided to incorporate on the off chance that my idea hit the big time. Although I am very interested in making my money back, my aim is to create learning exercises that generations can enjoy.

Setbacks:

At first, as a novice, I used Elance. That should be enough for even the most accomplished developers to cringe. I thought because I was an independent entrepreneur, I could find a like programmer to share in the discovery and excitement of building an idea together with me. I was bilked out of a large sum from a poor programmer. Elance was hardly helpful in the resolution even though they would say they were. I'm a teacher at heart who has a compassionate soul to help others. I learned early on that running a business requires a different mindset when people you pay make mistakes. Fortunately, I was eventually able to partner with an accomplished programming team, and we continue to bring more applications ideas to life. My message to anyone interested in creating an app: do your research; ask for references and get out when you feel something isn't going as you expect it should.

Predictable

FREE
By **Therapy Box**

Description

Predictable is an exciting text-to-speech application for the iPad, iPhone and iPod Touch. Offering customizable AAC functions with the latest social media integration, Predictable sets a new benchmark. Using a word prediction engine and switch access, Predictable meets the needs of a wide range of people using AAC. A wide range of people are using Predictable, including those with MND / ALS, Cerebral Palsy and people with communication difficulties after a stroke or head injury.

Company Profile:

Therapy Box was founded in 2010 by speech and language therapist Rebecca Bright and telecoms expert Swapnil Gadgil. It is a business that specializes in communication and therapy apps for people of all ages and abilities.

Therapy Box seeks to apply the latest innovations in app development to meet the needs of those who have communication difficulties due to a range of disabilities arising from cerebral palsy, motor neurone disease, autism, Brain injuries and other neurological and developmental disorders. The company aims to supply the best means of communication possible to our customers to enable them to express themselves with their family, friends and colleagues in a whole range of environments.

Therapy Box

Website:
www.tboxapps.com

Email:
info@tboxapps.com

Awards:

- 2011: mPhasis Universal Design Award
- 2012: Shortlisted – Learning Without Frontiers Award
- 2012: Shortlisted – Appsters Awards – Best Consumer App
- 2012: Shortlisted – Technology4Good Awards
- 2012: Shortlisted – Education Investor Awards

"The Making Of" Story:

Several years ago, Therapy Box Director, Rebecca Bright was looking for an augmentative and alternative communication (AAC) device to help her grandmother, who had motor neurone disease (also known as amyotrophic lateral sclerosis/ALS or Lou Gehrig's disease). Rebecca wanted a device that was affordable, easy to use and socially acceptable. She wasn't able to find a perfect solution. In 2010, Rebecca — who is a speech and language pathologist — envisioned a solution that did not yet exist. She chose to utilize the iOS platform and create her own AAC device. She developed the structure of an app called Predictable and worked with a team of skilled programmers and designers to code the app — which she knew would help many people with special needs, like her grandmother.

Successes:

Predictable is a life changing, sophisticated communication aid app that helps people to communicate with those around them. Users simply key in their message and press 'Speak'. We have received feedback from many of the thousands of users of the app. One example of a customer using it successfully is Kerry.

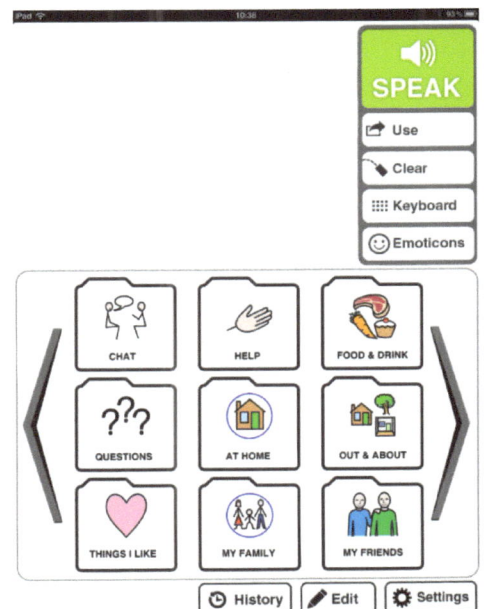

Kerry was diagnosed with MND in 2010, this lead to difficulties with his speech and the need for augmentative and alternative communication (AAC). His speech problems started with slurring. Kerry found people were unable to understand him over the phone, or in meetings. This had a big impact, and he could no longer talk to his wife of 37 years, Kathy, to explain his basic needs or to his family or to the careers and therapy team which was essential in his day to day routine. Kerry reports that using Predictable has allowed him greater ability to communicate – that it has given him a voice.

Predictable (English) frequently features in the top 5 grossing education iPad/iPhone apps for the UK, Australia, Ireland and New Zealand. Predictable (Dansk) is also a top grossing education app in Denmark.

Tips and Secrets:
Therapy Box's team of experts and developers focus on people who have communication difficulties due to a range of disabilities arising from cerebral palsy, motor neurone disease, autism, brain injuries and other neurological and developmental disorders. Therapy Box considered how they would communicate with others and how to make this whole process as efficient and effortless as possible for them and their therapists, friends and family.

Understanding our customers and engaging with them directly, taking all negative feedback as an opportunity to review how the app works and considering the contact with our customers gave us a valuable insight and was one factor leading to the success of Predictable.

Marketing Techniques:
In marketing Predictable, as well as the other specialist apps created by Therapy Box, the use of social media and presence at key industry events is essential. Reviews by influential and interested bloggers and writers has also garnered further media coverage.

Partnership Opportunities:
Therapy Box is happy to accept new projects/app ideas, and currently provide app development services across iOS, web and Android platforms. The in - house team of experts will help with app development from end to end app scoping and delivery, with ongoing support. The team has skills covering project management, app development and platform support services and have successfully delivered iPad and iPhone applications to customers all over the world. Anybody who has an idea or inquiry should contact appdev@therapy-box.co.uk.

Risks and Challenges:

One of the biggest challenges Therapy Box faced in the development of Predictable was ensuring the app meets the needs of as many as possible, while maintaining a clear interface and straight forward user experience. Creating an app for people with significant communication disabilities who rely on this app means that testing and quality assurance needs to be 100%. There is no room for a less than perfect app.

Setbacks:

Introducing a new way of providing augmentative communication to a market which has been unchanged for many years and relied on custom built, expensive locked-down PC based products is one of the setbacks that was faced when Predictable was launched. This has been overcome with increased awareness of the consumer and the need for health and education settings to manage tighter budgets and look to using mainstream and popular technology along with apps, such as Predictable, which meet the specific needs of this special client group.

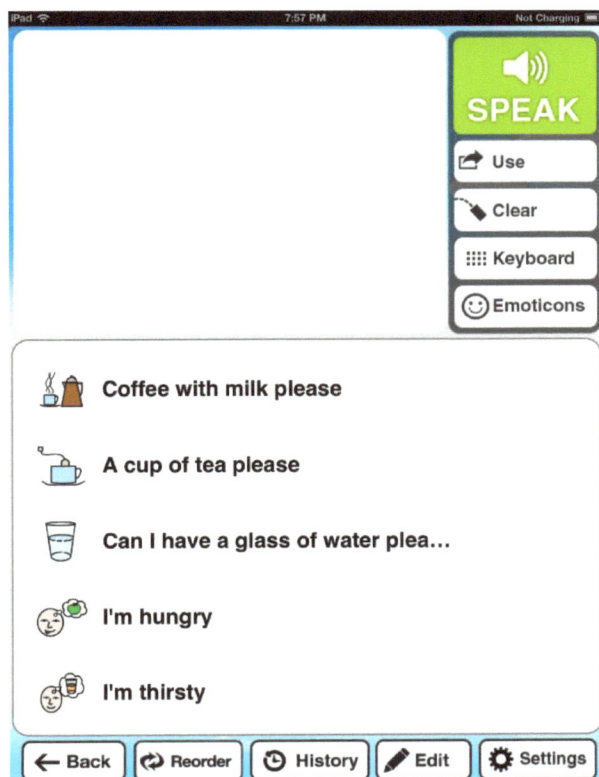

Smarty Print

$2.99
By SmartyShortz

Description

Smarty Print Handwriting in unlike any other letter tracing or handwriting app in the app store, but not unlike most SmartyShortz apps where there is always an element of fun and positive reinforcement!

Smarty Print allows a user to enjoy practicing and learning letters, write and read sight words then be rewarded with artistic creativity time that can be printed, emailed or saved to the device. There is never a better time to learn your letters than now with Smarty Print! Please use our Stylus Pencils with acclaimed grippers to help facilitate proper writing posture while tracing on the iPads.

Learn, Play, WRITE!

SmartyShortz

Website:
www.smartyshortz.com
Phone:
571-233-8626
Email:
jill@smartyshortz.com

Company Profile:

SmartyShortz has been developing quality educational children's apps since the first iPhone was released! Our staff at Smarty Shop outside of Washington DC, includes K-12 teachers, a resource teacher and gaming expert along with developers and designers. We created a Smarty Stylus Gripper Pencil to use on all touchscreen devices for children to continue developing fine motor skills while using swipe-intended devices.

"The Making Of" Story:

Smarty Print was created by a Kindergarten teacher needing an all-inclusive Language Arts App. She created it with teacher settings to be able to choose the font (Danelian or Zaner Bloser), the language (Spanish, English or Mandarin), any or all of the 300+ Dolch Sight Words or simply put in each child's name so they can practice writing it over and over. In the body of the app a user can trace upper and lower case letters, hear kids sounding them out, adults sounding them out as well as healthy foods that start with those letters. There is a teacher or parent assessment section to see where a child may have struggled with a letter and there are also art free time sections for users to do Jackson Pollock splatter paint, or create a coloring book from pictures on their devices or cartoons! A child can also record their voice saying words and then email grandma or take pictures in the app to turn into art! This app goes great with our Smarty Stylus Gripper Pencils mocked after a #2 pencil with the proper gripper, which she created for use with her iPads and Smart Tables where she was concerned swiping motion was altering her students' fine motor skills.

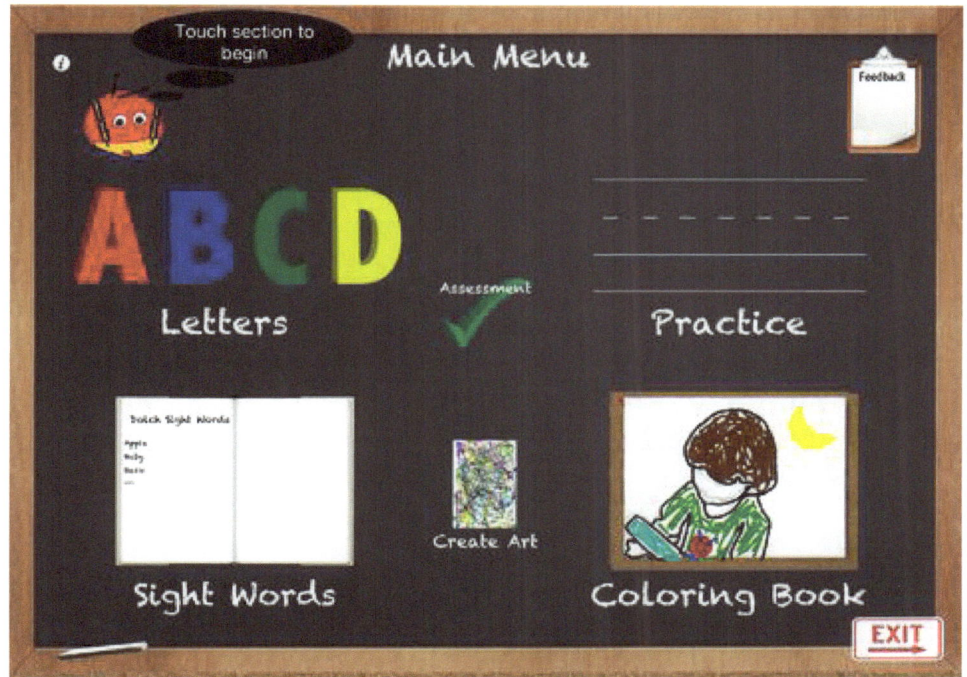

Successes:

This app was released in Nov 2012 and, out of our 15 apps, this one is week after week our top seller with our best day being over 800 downloads!

Tips and Secrets:

Creating an app with everything included. It's easier and faster to release an app with the individual components (small bits of what users want) but using several focus groups of teachers and adding multiple facets to the app make it a most desirable app which yields more downloads!

Marketing Techniques:
DAILY MAINTENANCE! It's a full time, daily job to be online promoting it through all channels.

Partnership Opportunities:
We have created over 15 of our own apps but also a dozen plus for creators looking to have their app make it to the app store! Please contact us if you are interested.

Setbacks:
Our first couple of apps were created by teachers but the graphics were created by kids themselves, and we got raked over coals in the reviews that our graphics were not up to "Angry Bird" status....it hurt the kids' feelings but we kept most of them.

iEvidence

FREE
By **Childcare Compliance**

Description

This app is designed for parents, educators and anyone else faced with an urgent situation involving children. If you have an emergency, call 911 or the appropriate authorities. We are not a law enforcement agency and this information goes nowhere but your email. If there has been a child-related incident you would like to document, this app will walk you through each step of the investigation. A preliminary report will be sent to your email.

User has read, understands and accepts the terms below before proceeding with purchase.

Profile:

Michelle McGinnis is a criminal prosecutor in Los Angeles and has worked for the prosecution for more than 17 years. Her experience includes school safety, gangs, narcotics enforcement, vehicle code enforcement, nuisance abatement, and animal cruelty and neglect.

Childcare Compliance

Website:
www.childcarecompliance.com
Phone:
213-637-0227
Email:
info@childcarecompliance.com

Prior to and during her legal career, Michelle worked as an educator holding teaching credentials and classroom experience in early childhood, K-12 and adult education. She has owned and operated childcare centers for more than 10 years and has spent a lifetime in the childcare industry. What distinguishes Michelle from many other lawyers is her experience as a childcare provider serving more than 200 families.

Moreover, in 2001, Michelle used her legal expertise and knowledge of the childcare industry to establish Childcare Compliance, now the nation's leading resource for legal compliance solutions available to childcare providers, agencies, unions, and parents. As executive director of Childcare Compliance (CCC), Michelle has guided a team of attorneys, educators, and IT professionals in perfecting legal and business solutions that protect schools and the children they serve by offering industry-leading documents and best practices for legal compliance across childcare, family, and labor law.

CCC has introduced iEvidence, a new mobile application designed to help anyone wishing to document an incident involving children in a school setting, including parents, educators and childcare providers. This can include bullying, child abuse, personal injuries, vandalism, and traffic incidents. Available from iTunes and Google Play for 99 cents, iEvidence is the only child safety mobile app that draws on the same court-related Preliminary Report used by police for collecting evidence from on-scene witnesses.

"The Making Of" Story:
Following a career as a teacher, I worked as a school safety prosecutor in Los Angeles for many years. Our work in school safety resulted in the publication of a school safety manual. After leaving a project in Watts, California and seeing firsthand how many cases go unreported simply because the average parent and student didn't know how to effectively report crime to law enforcement, I began to develop a mobile app. Using all of my savings, I developed iEvidence.

It is designed to answer the same questions victims and witnesses will be asked by police or on the witness stand. Using iEvidence is the one way to ensure crime is reported in a way that it cannot be ignored.

Successes:
A few students have used iEvidence and recorded testimonials indicating how they regret not having a tool like this when they were battling years of abuse and bullying by students and teachers.

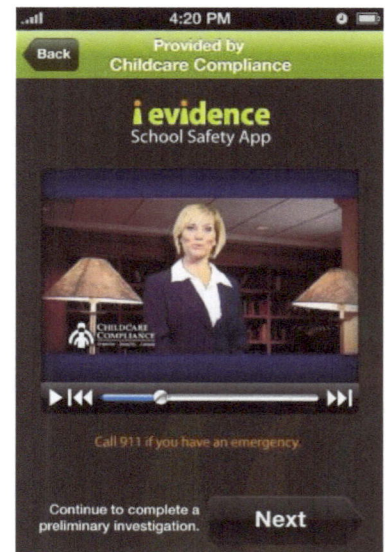

Risks and Challenges:

The biggest risk in being a non-tech person in a technical field is outsourcing and financing your first mobile app. I knew there was a need, but EVERYTHING else I had to learn from scratch. I lost a tremendous amount of money learning along the way, but would have never traded the learning process for anything. I can now call myself a developer instead of a silent investor. With the knowledge and relationships I have made throughout this experience, I can develop many more apps.

Setbacks:

Finding the right team of engineers and developers was the hardest lesson of them all. The major setbacks involved false starts with the wrong person. Being completely unfamiliar with the industry, I learned that only knowing the subject matter was a small and almost insignificant piece of the equation.

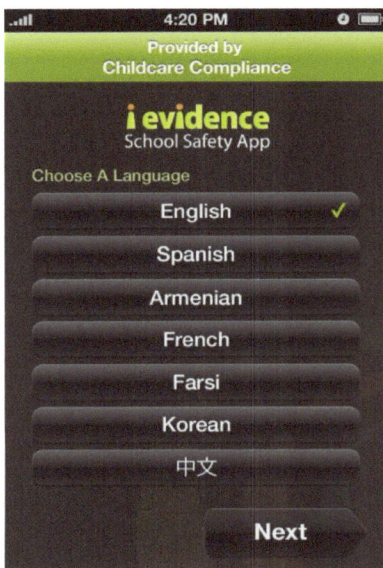

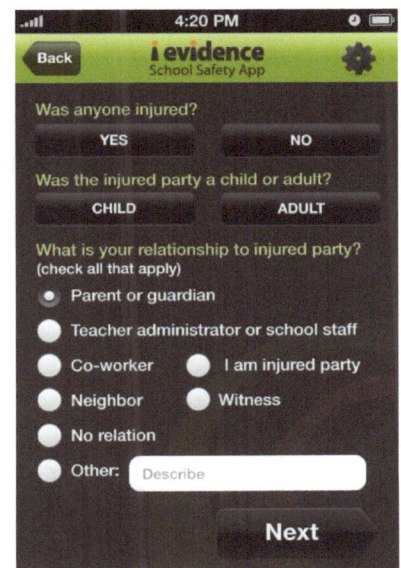

Distant Suns-Max:

Unleash Your Inner Astronaut

FREE
By First Light Design

Description

"Distant Suns is easy to use and understand. The graphics are amazing. I use it every chance I get. I love showing people the stars." Lee Brandon-Cremer, Space Shuttle Almanac

Distant Suns has been your personal guide to the cosmos since 1985. First on the desktop, now right in your hands. While others play games, you can travel to the stars.

Award-winning Distant Suns has the reputation of both having one of the most realistic displays of the night sky, while being one of the easiest to use astronomy apps for casual sky watchers as well as serious astronomers.

Features Include:

- Compatible with the new retina iPad.
- No Internet required. Just a sense of wonder.
- View from the Earth or out in the solar system.
- Observe the targets being monitored in real-time, by the SETI Institute for signs of possible extraterrestrial intelligence.

First Light Design

Website:
www.distantsuns.com
Email:
mike@distantsun.com
Phone:
408-383-0627

- Augmented reality viewing overlays the sky with your local landscape.
- Compass aware (iPhone/iPad only). Simply aim and gaze.
- Over 300,000 pinpoint stars scintillating like diamond dust in the palm of your hand .
- Integrated with NASA's Night Sky Network space events in your area (US only at present).
- What's Up? Gives a quick one snapshot overview of the evening's sky.
- Realistic ghostly band of the Milky Way.
- Stories behind each constellation.
- Viewpoint Lock keeps any planet centered.
- Current weather patterns on the earth, updated daily.
- Change the look of the stars.
- Galaxies, nebula and star clusters.
- Hubble Space Telescope images.
- All 8 planets (Pluto is optional).
- GPS aware.
- Special color mode preserves your night vision.
- Touch the sky to reveal hidden data for each object.
- Planetary data and information.

First Light is a proud member of Moms with Apps, a collaborative group of family-friendly developers with family-friendly apps.

Awards:
- Named in the "100 Essential Apps" by iLounge
- Featured on NPR's MarketPlace, October, 2010
- 5 Stars from ituneappreview.com,
- 5 Stars in iphoneapplicationlist.com

Galleries HQ

FREE
By **Arbutus Software Inc.**

Available on the App Store

Description

Beautifully designed for the iPad, Galleries HQ presents the world of art and architecture from locations around the world. Use the interactive map to find artists of interest, or search using our 20+ art-related categories.

Ideal for travelers, students and all art enthusiasts! Galleries HQ lets you explore the world to find your favorite artist. Need more information? Contact people directly using twitter, email or their website.

See why we're excited to bring the world of contemporary art to the iPad!

Arbutus Software Inc.

Website:
www.gallerieshq.com

E-mail:
hello@gallerieshq.com

"The Making Of" Story:

The idea behind Galleries HQ started in late 2009. During that time there was a lot of buzz in the community about developers creating innovative 'apps' that combined Internet connectivity with the capabilities of GPS enabled 'smart' phones. Being an avid art enthusiast and traveler, I thought about how I could use a device like the iPad to allow people to experience art and culture. Fast-forward to 2013, and there are numerous websites that promote art. With Galleries HQ I wanted to help tell a new story with an entirely different platform.

Successes:

For the app, we had to come up with our own unique design. This involved not only writing code, but talking directly with artists, attending exhibits, visiting museums, lectures and art fairs. This proved invaluable in many areas, including the design of our map-driven interface.

Tips and Secrets:
Even though we're known for our app, we consider ourselves be a service-based organization. For us this means building great relationships with artists and always thinking of technologies that may be useful. This goes well beyond the iPad and is an ongoing exercise.

Marketing Techniques:
Our marketing strategy focuses on building community through various face-to-face and digital strategies. Having the patience to test various channels is also important. In addition, we place great importance in our "company voice" and building trust with our audience.

Partnership Opportunities:
We'd love to hear from other art-based organizations interested in business-to-business (B2B) partnerships.

Risks & Challenges:
One assumption we see with other app companies is their sole reliance on The App Store for their marketing and revenue generation. While this has proven successful for some, companies should assume responsibility for their own products and think of ways to provide exceptional value. This realization has allowed us to build a much richer ecosystem than was originally planned.

Setbacks:
For our app, the biggest challenge was creating a search model that made sense. We had a number of ideas that used traditional paradigms but we felt we could do more. The result was using a world map as the main search and discovery interface. It took a lot of thinking but we're so pleased with the results!

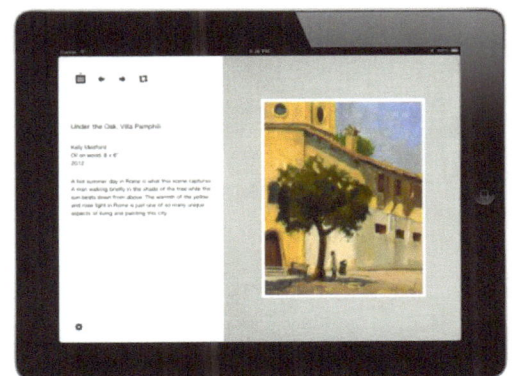

Chapter 14

Entertainment

"The length of a film should be directly related to the endurance of the human bladder."
~ Alfred Hitchcock

9x9.tv

FREE
By **9x9.tv**

Description

9x9 app instantly turns your Android device into an Android TV starring your favorite YouTube characters. With 9x9 app, there is always something good to watch on your Android any time of the day. Begin with 9 fresh YouTube channels carefully curated according to the time of the day or dive deep into a variety of channels that cater to your unique lifestyle. Flip through curated YouTube channels or playlists at your own pace just like watching TV. Browse the next episode in a channel or jump to another channel with just the swipe of a finger.

Tune in at any time for a personalized bite-sized selection of channels with content that informs and entertains. For example, the "Morning Shows" time slot provides a channel line-up consisting of breaking news and morning talk show channels. "Late Night" on 9x9 features comedy and late night talk shows. So just sit back and let 9x9 entertain you with one episode after another non-stop. Turn on the excitement with every flip as you navigate the channels prepared for you by 9x9's curators to discover Internet videos you never knew existed. Let 9x9 guide you through a personalized video tour of what matters to you every time you flip it on.

9x9.tv

Website:
www.9x9.tv
Phone:
408.970.3318
Email:
dan.lee@9x9.tv

Company Profile:

9x9.tv is a Silicon Valley company headquartered in Santa Clara, California with international offices in Taipei and Beijing. Funded by venture capitalists, private investors and corporate investors, 9x9.tv is leading the way to revolutionize how Web videos are managed, delivered and consumed.

"The Making Of" Story:

At 9x9.tv, we've always wanted to curate and bring the near-infinite amount of web videos to those who are used to watching TV. Therefore, integrating curated web video content with traditional TV-watching behavior has been one of our top priorities. With that in mind, we wanted to bring these videos to our users in a way to which they're accustomed. During the planning phase, we also listened to a lot of user feedback and took cues from traditional broadcasting techniques. The result is a revolutionary way to watch web videos in the same fashion as watching TV. 9x9.tv is one of the first Android apps to allow a user to "flip channels" with the swipe of a finger. On the mobile platform, some people call us "Flipboard for videos". On the TV platform, 9x9.tv offers users a lean-back experience to watch web videos curated into TV-like channels which can be easily surfed like traditional TV.

Successes:

The 9x9.tv Android app achieved approximately 5,000 downloads within the first five weeks of launching.

Tips and Secrets:

We're able to be successful by adhering to the principle that "less is more" and by giving the users what they want in consuming and sharing web videos through easy and intuitive user interfaces across all screens.

Marketing Techniques:

In preparation for the launch of the 9x9.tv Android app, we targeted key media influencers such as *TechCrunch* and *The Next Web* who showcased 9x9.tv as a premier app using YouTube's just released Android API. We also reached out to publications that specifically focus on mobile applications such as *Appolicious*, *Android Police*, and *Android Central*. These outlets featured 9x9.tv in editorial reviews, which heightened awareness of both the brand and product, and drove downloads.

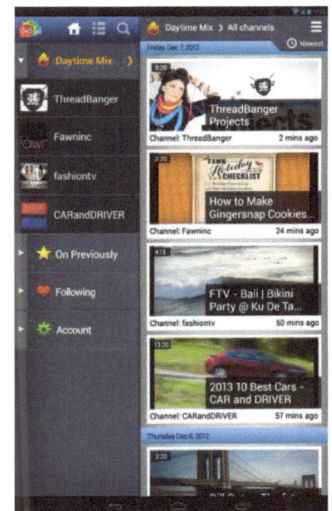

Partnership Opportunities:

We welcome all inquiries! Our passion is to define the next "video internet". The 9x9.tv team members are all geeks at heart. Nothing's better to us then discussing exciting new ideas with those who also enjoy pushing the envelope.

Risks and Challenges:

The hyper-abundant supply of videos on the internet is no longer our main challenge. Rather, our biggest concern is the users' attention span. As a small team, we constantly have to pivot with the environment and experiment with technology to determine how to best capture users' attention span, and how to do it fast. 9x9.tv is able to overcome this hurdle through a combination of algorithm and humanism in order to deliver radically friendly solutions in an established ecosystem such as the TV industry. For example, 9x9.tv has an in-house team of curators assisted by machine-generated algorithms to bring the most relevant video content channelized for a specific target audience.

Setbacks:

Finding the most relevant web videos to present to our users is a constantly evolving process simply because "relevance" means something different to everyone and a user's lifestyle may evolve. Before we began recommending channels by time (also known as "day parting") on our Android app, 9x9.tv tried various other recommendation strategies such as popularity and social relationships between videos, resulting in minimal differentiation and user traction. Additionally, these tactics didn't help us move closer to traditional TV screens. Now that we've changed our thinking and are recommending fresh web videos and pushing users' existing web video subscriptions (e.g., YouTube channels) based on the time-of-day, our users have a reason to turn to 9x9.tv first at any time of the day to enjoy video content.

Hollywood Camera

$0.99
By **Ribui**

Description

Amaze and delight your friends and family by dressing up in the actual costumes from blockbuster Hollywood movies! Have you ever wanted to know what you would look like dressed as Maximus from Gladiator, Queen Elizabeth I or the Blues Brothers? Hollywood Camera is the only app that instantly turns you into a movie star and enables you to wear the actual costumes that have been immortalized by some of the most iconic film characters from a century of Hollywood filmmaking.

Launched in conjunction with the Victoria and Albert Museum's Hollywood Costume exhibition, there are 16 costumes to choose from including Charlie Chaplin's Tramp suit and Viola de Lesseps stunning dress, headdress and jewelry from Shakespeare in Love as worn by Gwyneth Paltrow. You can also adorn your costume with essential accessories: from the Blues Brothers' cool ensemble with the all-important sunglasses and hat to the combat armor of the Gladiator's Maximus worn by Russell Crowe. Hollywood Camera is beautifully simple to use - select a costume in the app, align the costume on a friend or front facing on yourself and snap the photograph. Then share the photo via email or social networks.

Ribui

Website:
www.ribui.com

Email:
hello@ribui.com

This is the most fun that you'll have with someone else's clothes on!

Company Profile:

Ribui is a UK-based leading digital publisher, specializing in the creation of augmented reality and 3D apps and interactive books for smartphones and tablets.

"The Making Of" Story:

We partnered with the Victoria & Albert museum to produce an application that would offer an additional way for the audience of the Hollywood Costumes Exhibition to engage with the famous outfits, and a way for people living far away from London to enjoy part of the show anyway. We wanted something both funny and informative, that people would really enjoy. The app works with a camera overlay system that allows users to virtually dress up with one of 16 official costumes from iconic films including The Blues Brothers, Shakespeare in Love Mamma Mia. We thought it would be fun to be able to use the front-facing camera and take a self-picture as well as be able to take a portrait of friends and family and then share the images with them. The production of Hollywood Camera lasted a few months during which we also had the chance to touch and feel the costumes (that's not allowed to visitors) while taking the pictures for the app and to learn a lot about movies and costume design when sourcing the information for the costume tables.

Successes:

Hollywood Camera ranked high in the Entertainment section in the UK, Canada, Australia, Ireland and Spain, also reaching the top 10 grossing in Malta.

Marketing Techniques:

We employed a specialist app PR and marketing agency, Dimoso, to launch the application. The marketing and communication team of the V&A Museum helped us spread the word on their channels and we scheduled a promotion with the app going free for a week-end to get visibility.

Risks and Challenges:

When we started working on this project we didn't know how long and difficult it could be to turn the picture of costume into something that can be overlaid to a person's photo without looking artificial and cheap and we focused on getting the best result, spending time and resources on it. We finally obtained a great result, thanks to our team's great professionalism.

Setbacks:

We only implemented the feature to upload a picture from the camera after an update of the application, given the feedback of the users. If we had more time, we would have launched Hollywood Camera with the feature included, but we had a strict deadline to respect: the opening day of the exhibition. For future projects, we'll try to take the project to the level we want and only launch when we're happy with it – the app store users have high expectations from paid apps.

Movie Spots

$1.99
By **Christopher Warner**

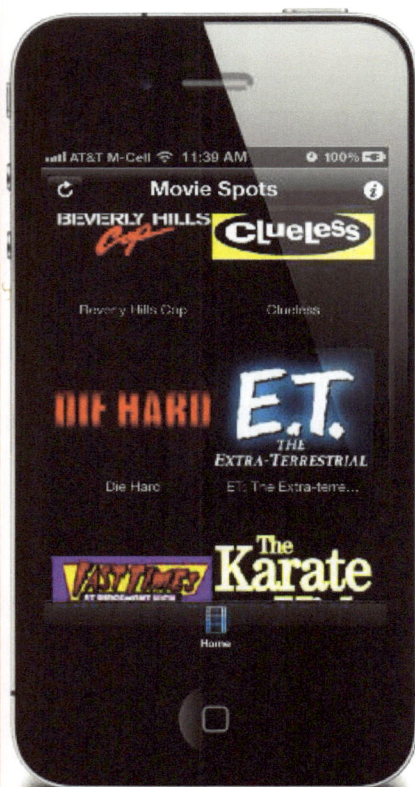

Description

The ultimate travel guide to Hollywood film locations!

With movie tourism now a global phenomenon, here you'll find original movie spots to some of your favorite movies.

If you ever wondered 'Where did they film that?', or you want to visit the filming locations of your favorite movie, simply navigate the app via movie title and you are on your way!

Christopher Warner

Website:
www.hollywoodfamine.com

Email:
info@hollywoodfamine.com

Chapter 15

Games

"I went to a fight the other night, and a hockey game broke out." ~Rodney Dangerfield

Real Boxing™

FREE
By **Vivid Games**

Available on the App Store

Description

Real Boxing™ brings you the most exhilarating, no holds barred fighting action ever seen on the App Store, taking you closer to the action than you ever thought possible. Combining eye-popping ultra-realistic graphics with the revolutionary V-Motion Gesture Control System, Real Boxing lets you control your fighter with real punches through your device's camera.

Featuring ultra-realistic motion capture from real boxers plus amazing graphics from the powerful Unreal Engine 3, you'll see sweat and blood fill the air and you'll feel every hook, jab, and uppercut in Real Boxing, the most exciting boxing game yet on the App Store.

Vivid Games

Website:
www.vividgames.com

Email:
info@vividgames.com

Company Profile:

Vivid Games is one of Europe's premier emerging independent studios, with a passion for accessible and engaging gaming at the heart of its philosophy. The company is built on a solid commitment to excellence in all aspects of game development, balancing design, innovation and cross-platform technology. It believes in user-focused playability and a strong awareness of the ever changing trends in modern gaming.

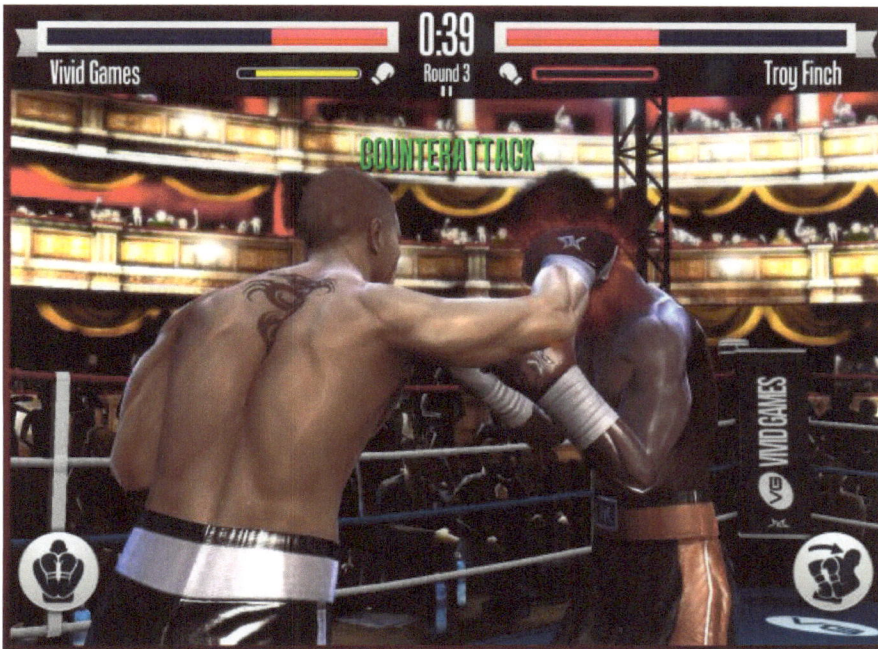

Founded in 2006 in the heart of Poland, Vivid Games is making a valuable contribution to the worldwide games development community. Having already expanded its presence to London, England, the company is also spreading across the Atlantic to San Francisco in the United States.

Since its inception, the release of over 150 titles has seen Vivid Games grow to become a globally recognized force in the realm of mobile and digital games. Often working in long-term partnerships with some of the world's leading game companies, Vivid Games has become renowned for its specialization in cross-platform development, working both with licensed brands and exciting new IPs of its own.

"The Making Of" Story:

We wanted to produce a AAA title for iOS/iPad and while deciding on our next game, we saw a great opportunity to bring a realistic boxing title to mobiles and tablets. With EA's Fight Night Champion being the only real contender for the boxing title, we knew that we would "enter the ring" as the wildcard outsider. Still, from the very beginning we wanted a swing at the championship belt.

The production of Real Boxing™ lasted 5 months and involved 30+ developers. We had motion capture sessions with professional boxers.

Successes:

Real Boxing was featured by Apple as Editor's Choice, and reached the number 1 spot in more than 70 countries, being in the top ten of 131 app stores. Real Boxing was also featured in New and Noteworthy, Action Games, Sport Games, Games for Your New iPhone, What's Hot, Cutting Edge Games, and App Store Selection 2012.

Real Boxing was also chosen by NVIDIA to be optimized for their Shield device and was showcased at their keynote at CES 2013.

Tips and Secrets:

Real Boxing has by far and away been the biggest venture for our studio and brought invaluable experience. We focused on the game for months and fought to find solutions to some technical problems, asking the team to work harder and harder. The secret to Real Boxing's success has been the great involvement and professionalism of all our team members, who we can't stop to thank.

Marketing Techniques:

We employed a specialist app PR and marketing agency, Dimoso, for the launch of all our titles. For Real Boxing they planned a 6-months-long strategy to create hype around the launch and bring attention to the best features. They distributed exclusive content to the media and made sure that journalists would review the game and publish their articles in November, on launch day. To keep the momentum and stay high on the App Store charts, we offered a limited-time discount using Free App a Day, and published a big update with new content in December.

Partnership Opportunities:

We're accepting developer projects.

Setbacks:

Despite hours and hours of testing, a few nasty bugs afflicted the launch, with the lower-end devices experiencing some problems during the game tutorial. Thanks to users' feedback, we fixed all of them, but we paid the price by getting some negative reviews on the app store.

Chapter 16

Photo and Video

"A true photograph need not be explained, nor can it be contained in words." ~ *Ansel Adams*

Cycloramic

$0.99
By **Egos Ventures Inc.**

Description

Cycloramic brings you a whole new way to take panoramic pictures. It lets you take panoramic photos and videos so easily, you can do it with your eyes closed (with iPhone 4/4S/5) or hands free (with iPhone 5 only)

Company Profile:

Atlanta-based Egos Ventures is a privately held innovation lab with projects in the Big Data and mobile sectors. The company has developed an intelligent recommendation platform that leverages location-based and behavioral data. The critically acclaimed Cycloramic is the company's first entry in the mobile app sector. The app enjoyed more than 100,000 downloads shortly after its availability in the App Store in December 21, 2012.

Awards:

- The New York Times: Cycloramic gets the Pogie Award for the Brightest Ideas of 2012. (12/17/12)
- TechCrunch: Steve Wozniak Uses Cycloramic and an iPhone 5 To Street View His Kitchen. (12/27/12)
- Digital Trends: New iPhone 5 app takes hands free 360-degree video. (12/31/12)
- ABC News: App of the Week: Cycloramic (12/29/12)
- GIZMODO: Woz Totally Loves Cycloramic (12/29/12

Egos Ventures

Website:
http://cycloramic.com
Phone:
404 643 2220
Email:
bruno@egosventures.com

- <u>The Verge</u>: Cycloramic app takes panoramas by vibrating your iPhone in a circle (12/28/12)
- <u>Los Angeles Times</u>: Cycloramic app spins iPhone 5 to take 360-degree video (12/27/12)
- <u>CNET</u>: How to make your iPhone 5 'dance' (12/27/12)
- <u>TUAW</u>: Cycloramic spins your iPhone 5 by itself (12/21/12)

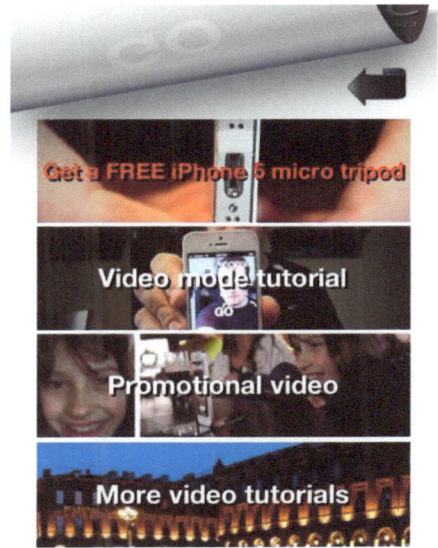

"The Making Of" Story:

We were getting toward the end of the year at Egos Ventures, and we had been working for months on our recommendation engine. One Friday afternoon, we wanted to take a break and decided that we were going to develop and launch an app before Christmas. The app would have to be different from anything in the app store. We realized it will be a challenge but thought that surely we could think of something that nobody had already thought about.

Boy, we were wrong! Every single crazy idea that we considered so unique that nobody could possibly have thought of it, was not only there but also already copied 20 times. After being disappointed about 20 times, we realized that this was not going well and that we would have to try a totally different approach. We then decided to look at the iPhone as a physical object rather than a computing device, so instead of going through a creative process to find an app, we followed a methodical approach. We looked at the phone as an object with many functionalities (screen, flash, front/back camera, speaker, microphone, compass, accelerometer...) and physical attributes (straight metal edges, glass front, metal back, rounded corners, buttons, speaker and connector holes...) Once we had that list for each functionality and characteristic, we listed every application that could be derived or hacked, including applications that were not meant for the original purpose (use a speaker as a microphone, use the flash as a code communication, use the vibration to move the phone, use the straight edge as a ruler...). Since we wanted to make sure nobody had thought about whatever it was that we were going to make, the last requirement was to combine at least two of the discovered applications. We came up with 6 apps, some don't exist yet but don't have much use. Some we may launch one day. I really liked the vibration moving the phone; I thought that would definitely add a new dimension. The problem is that the movement was uncontrolled and inconsistent depending on how and where the phone was placed. So we made a test app to be able to control different vibration patterns on commonly used surfaces and manually tried different patterns on different surfaces, while timing to find the optimum vibration switch frequency to be most consistent on most surfaces. As soon as I saw the phone rotating 360 degrees, I knew that we had to film and take pictures! We immediately searched the app store, and bingo, nobody had done it! Unbelievable! With only a short time before Christmas, we wrote it and submitted it to the app store. Apple rejected our metadata because they thought we were misleading users into believing that the phone will move by itself. Exactly! So we filmed the phone in action and sent it to Apple, this was the last day to be approved before the app store closed for the holidays. Within a few days the media started to talk about it and we won the Pogie Award for the "Brightest Idea of 2012." The magic happened when the Woz, Steve Wozniak, thanked me for the app on Christmas night and later sent me a video of himself using the app in his kitchen.

Marketing Technique:
The key to spreading the word about your app is through articles on technology websites and blogs. It takes a lot of time to compile a strong list of contacts with their email addresses. They don't readily give their contact information away, so it takes time to track them down. Websites like Muckrack and Journalisted can help. Make sure that you have something newsworthy and exciting to tell them. Don't write a lengthy press release but send them a short snappy message. Further, try to be personal, comment on a relevant article that they have written and don't just cut and paste to everyone.

One of our priorities has been to have high quality promotional videos to post on YouTube and Facebook page. This is also a great tool to send to journalists.

Setbacks:
We had a miscommunication with one of our media contacts regarding the launch of version 2. They printed that the app was available before we were ready to launch. It was the journalist with the largest readership and so we decided to launch the app in line with the article. We were not entirely ready to launch and did not have our promotional video in place.

Moshpic

FREE
By Maximilian & Co.

Available on the App Store

Description
Moshpic is the best way to privately share photos with friends and family.

Moshpic lets you share stacks of photos between iPhones & effortlessly keeps them synced. It's an intimate photo sharing experience like no other.

Oh yeah... it's completely FREE.

"The Making Of" Story:
As they say, necessity is the mother of invention. We made this app as an answer to a big problem we had, which was being able to effortlessly send pictures from device to device. Once we got a stable build and were able to see our idea come to life, we realized we had something very useful.

Maximilian & Co.

Website:
http://moshpic.com
Phone:
818-658-7137
Email:
hello@moshpic.com

Tips and Secrets:
We are far from perfect, but we do try to spend a lot of time on the details. We are also very willing to totally trash weeks of work if it means the user experience will benefit.

Marketing Techniques:
At this point, we pretty much only have time to work on designing and developing the app. For now, our marketing technique ends at nice visuals and simple copy.

Risks and Challenges:

Obviously spending your own time, money, and effort is the risk we face here. Is it worth it to do this? Are we putting something useful into the world? We hope so.

Setbacks:

Tech moves fast. We're willing to spend time on little things and make sure things are right before taking the next step, but sometimes it's difficult to keep that steely resolve in the face of all the stories you read every day. You're mostly led to believe you're not working hard or fast enough.

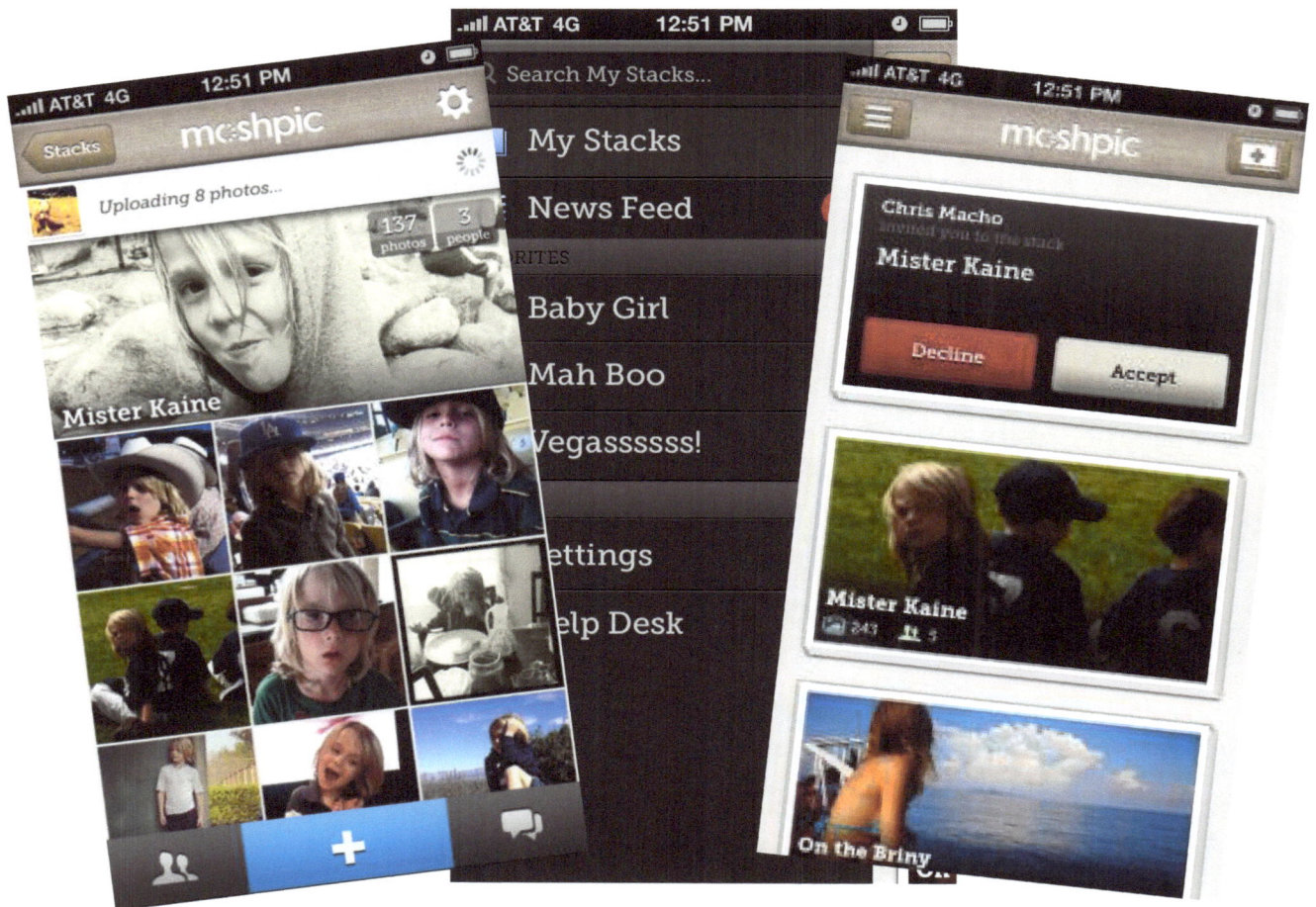

PicsArt Photo Studio

FREE
By PicsArt

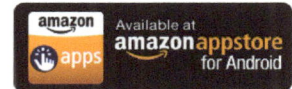

App Store · Google play · amazon appstore for Android

Description

PicsArt is a free, full-featured photo editor and art community that was born out of a sheer frustration that no free photo app seemed to have a robust feature set. When it launched in November 2011, PicsArt remedied this problem and today, PicsArt dominates the Android and Kindle markets for photography. PicsArt delivers more than crop and filter functions – it packs a full art studio with professional-grade tools into an easy and fun-to-use app. Users can add "motion-blur" and "blending" effects, create collages, and insert text, clipart and more. PicsArt consistently ranks in the top 50 free apps from Google's Play store, has more than 45 million downloads and is growing at more than 3 million downloads per month. With PicsArt's launch on iOS in January, the app's development has only just begun.

PicsArt

Website:
http://picsart.com
Phone:
(550) 877-2952
Email:
art@picsart.com

"The Making Of" Story:

We started as a team in Armenia that developed imaging apps for Facebook and MySpace back in 2007. With the advent of Android, we sense an opportunity and began to develop several different image and photo-related apps that were all focused on single benefit functionality. By November 2011 we had decided that what users really wanted was not a bunch of different apps, but a single, free, full-featured photo app that allowed them to create and transform imagery all within one ecosystem. No one else was doing this at the time and PicsArt was born. Two months later we were already the #1 photo app on Google and we've held that position ever since.

We've achieved this without being funded and without ever being promoted in any way by Google.

Successes:

PicsArt consistently ranks in the top 50 free apps from Google's Play store, has more than 45 million downloads and is growing at more than 3 million downloads per month.

Tips and Secrets:

What makes PicsArt so successful is that we are a community-driven app. Our development team is constantly tuned in to what our users are saying – what they like, don't like, and what they want to have. PicsArt isn't our app. It belongs to our users. This relationship we have with our users has been the key to our success.

Partnership Opportunities:

We are always open to listening to interesting ideas.

Risks and Challenges:

When we first started, Android was far from a ready platform and as we've grown we've had to support the full spectrum of Android users from Android 2 to Android 4.2.1. This was and remains a massive challenge, even as Android has massively improved as an OS. Add to this, the over 1,400 models of Android-based smart phones and keeping our users up and running across such a diverse environment is our biggest challenge.

In terms of risks, based on our early success, we were pushed right away to move into iPhone. But we chose to wait and build soul into our Android app. Once we felt like we had an app that really reflected what mobile photographers wanted we felt we were ready to launch in iOS, which we did in January of this year. Had we launched in iOS earlier, we could have seriously damaged our business because we would have been spread too thin too quickly.

Tapshare

FREE
By **Tapshare**

Available on the App Store

Description

The easiest way to make & share photo albums! Multiple people can add to a photo album - group photo albums!

Create an album for your first date. You can both add to the album! Having a holiday party? Everyone can add to the photo album! Tapshare is perfect for sporting events - the whole team can add! Tapshare is perfect for wedding photo albums! Everyone's photos, all in one place. Finally!

Tapshare is the easiest way to share a lot of photos. Tapshare creates instant collages for you to share to Facebook and Twitter.

YOLO! Tapshare it.

Company Profile:
Tapshare is a Los Angeles based company. The app lets groups of people add photos to a shared album. Tapshare was started by two friends who met in High School and grew up with the Internet. It was built with no budget in their apartment in North Hollywood.

Tapshare

Website:
tapshare.com

Email:
contact@tapshare.com

"The Making Of" Story:

In the early 2000's, one of our High School friends showed us a new digital camera he got. It was the size of a Polaroid camera and you inserted a floppy disc as the memory. It was brilliant. Digital cameras rapidly started shrinking. Around five years later they could fit in a pocket. We each got our own digital cameras and took tens of thousands of photos everywhere we went. Another five years later, we got iPhones, which included a good quality camera with internet connectivity. We felt it was the perfect opportunity to create an app which gathered the photos we take together into a single place. One year later, Tapshare was born.

Successes:

Tapshare has helped capture thousands of memories from groups of people around the world. A room full of people adding photos to a shared album captures a shared experience that would have been lost forever.

Tips and Secrets:

Listen to your customer, see how people use it, constantly improve, and most of all, be your own #1 power user.

Marketing Techniques:

Don't be afraid to get your product out there. We gave a beta version to our friends before it was finished. They gave us incredibly useful feedback and when we finally launched, our friends told their friends, who told their friends, etc...

Risks and Challenges:

We turned down high paying jobs to bootstrap Tapshare to success. We put it all on the line and went all in.

Setbacks:

We pivoted a couple times in the beginning and by getting feedback early on, we were able to quickly change the product.

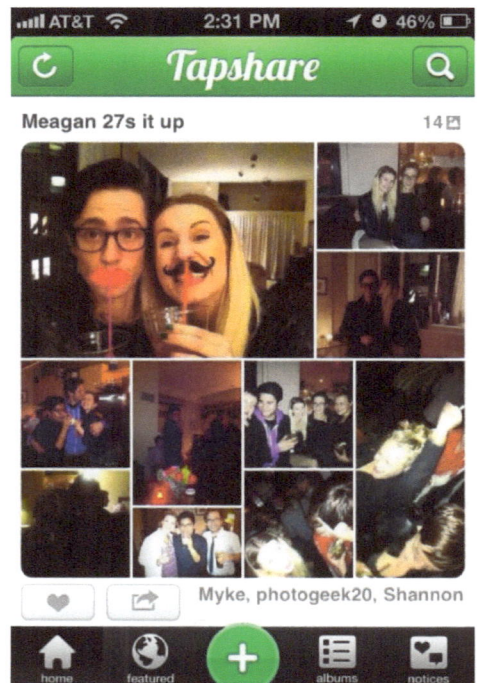

Chapter 17

Social Networking

"On Twitter we get excited if someone follows us. In real life we get really scared and run away."
~ *Anonymous*

Free Safe Text

FREE
By **3i Tech Works, Inc**

Description

The app works with the GPS in your phone. As you reach 10 to 15 miles per hour it becomes active or you can turn it on manually (push ON or shake the phone) at any time you wish to be in FreeSafeText mode.

When the app is on, as you receive a message your phone sends out an automatic reply that says, "I received your message but I am in FreeSafeText mode. I will get back to you asap."

Company profile:

3i Tech Works, Inc. stands for Implementing Innovative Ideas, an incubator that implements ideas that vastly improve the way we live, work and play. We are located in Boca Raton, Florida, next to the Florida Atlantic University campus. A company dedicated to the development and installation of mobile apps that add significant value for its user base.

"The Making Of" Story:

We found while driving that everyone was distracted while texting and weaving through traffic. I had several instances where I had my one year old grandson in the car and someone texting and driving almost hit us. I was inspired to not only help him, but all families.

3i Tech Works

Website:
www.FreeSafeText.com
Phone:
561-244-9490
Email:
joer@3itechworks.com

Having received a text, I thought it would be a good idea to have an auto responder to let others know that you are driving. I wanted to make the app universal for all to use and have sponsors pay the tab for our users.

Successes:

I enlisted legendary Coach Howard Schnellenberger to tell our story. In less than 4 months of a PR campaign, we have been covered by radio, TV, magazines and newspapers across the country. The personal story of the application hits close to home and is such a key issue that we have received more than 100 media placements during that time.

Marketing Strategies:

We are still telling others why this is a great app and are enlisting the help of major brands who support our idea of making driving safer.

We use Public Relations and appear on TV, radio, magazines and newspapers to spread the word. Since this is a personal story it connects with the lay technology user and the serious app user. Our PR firm believes that every business and every app has a story.

Partnership Opportunities:

We are always searching for new innovative ideas to implement; our readers can contact us via email at business@3itechworks.com for business development opportunities.

Risks and Challenges:

Getting the word out was the first challenge and then finding sponsors who believe in our project and team to move the application ahead of everyone else in this space.

Everloop Goobit

FREE
By Everloop

Description

The first safe, fun-filled social app for Everloop kids! It's the coolest way for kids to post, share photos, and prank (Goob) friends on the go. Everloop Goobit makes it safer, easier, and more fun to stay in touch with friends on the iPhone, iPad, & iPod touch.

- Post messages, photos, and videos, and share with friends
- Join fun loops about things kids love - music, sports, you name it
- Comment on friends' posts, pictures, and videos
- Like things
- Goob like crazy
- Stay safe with state-of-the-art technology to stop bullying and bad language
- Goobit supports iOS6 and is now faster and easier to use

Use Goobit while you're on the go…

Everloop

Website:
www.everloop.com

Phone:
(212) 580-0835

Company Profile:

Everloop.com provides a safe, privacy protected and content moderated social online platform for children 13 and under to embrace the positive opportunities of the Internet and develop core skills needed for the 21st century child including creativity, collaboration and communication. The site was co-founded in 2011 by three female entrepreneurs, Hilary DeCesare, Paige McCullough, and Kim Bruce.

Awards:
- DEMOgod Award for Social Media
- AlwaysOn Global and AlwaysOn Hollywood 100 Awards
- Parents' Choice Award
- Edison Award
- Golden Bridge Award's Silver Award for Social Media, Networking and Collaboration Innovations.

"The Making Of" Story:
Everloop is a social network for over 300,000 kids. In 2012, Everloop kids who got iPads, iPods, and iPhones asked for their own Everloop App. We decided to make our most fun feature, "goobing," mobile and support social check-in from devices. Using Goobit, kids send and receive goobs, animated pranks, from each other while keeping current with their social network and sharing photos and videos.

Successes:
Everloop Goobit is a free app for Everloop kids. 80% of active Goobit users sign in each day. While designed as an add-on for Everloop website kids, many Goobit users are App-only. Since its release in September, kids have signed in to Goobit over 9,000 times, interacted with Everloop over 70,000 times, made over 8,000 posts and comments, and sent almost 5,000 goobs.

Tips and Secrets:
Our approach to get Goobit adopted was to make a fun App, release it to our Everloop kids as a sneak peek, and then continually improve the App based on feedback from these kids.

Marketing Techniques:
Everloop Goobit has been publicized and promoted as the first safe social mobile app just for kids.

Risks and Challenges:
Everloop is dedicated to protect children by following

COPPA guidelines. Much interaction with Goobit needs to go through our filters and moderation. The greatest challenges we faced were in making the app both safe and fun.

Setbacks:

Everloop Goobit was the first application to use the new Everloop API. Since there was no API at the beginning of the project, we developed a prototype of Goobit using the well-known Twitter API. Once enough APIs were written, we ported Goobit but had to make extensive changes to ensure that every server interaction complied with our COPPA-driven business rules.

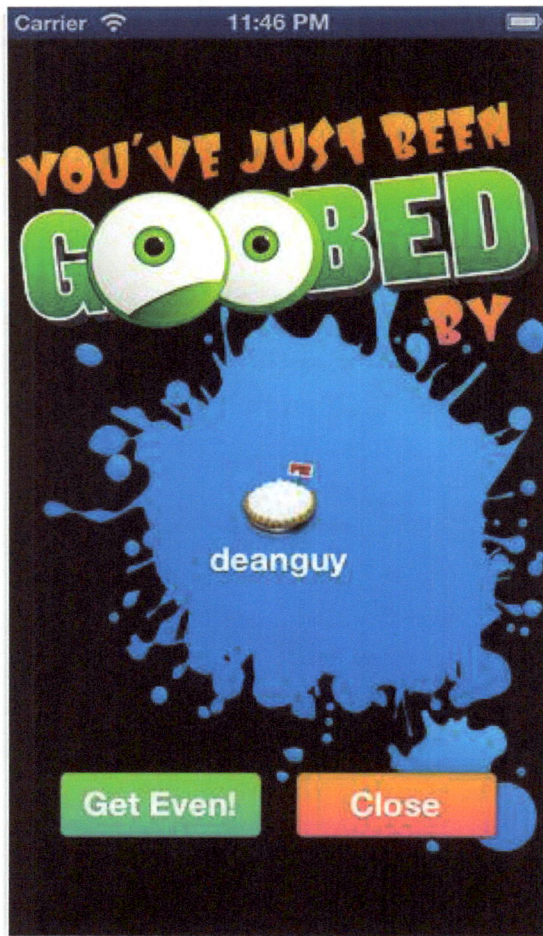

YappBox

FREE
By **Yapp Inc**

Description
YappBox is your inbox for all of your Yapps and those you have been invited to.

Yapp is a simple and fun way for consumers to create beautiful, content- centric mobile apps ("Yapps") around parts of their lives that matter to them. Yapps are user-created, themed, customizable, mobile experiences that can be published and updated in real time and do not require their own binary.

Company Profile:
Yapp has seen great recognition from the press since its launch, and has been featured in technology and mainstream media. We demoed the app live to the nation on Good Morning America, have appeared on CNN, and in major newspapers and blogs. The WSJ called the app "user-friendly," "beautiful," and "elegant;" Mashable voted us a stand-out app for 2012; The Huffington Post said we were one of four startups "changing the mobile space;" BizBash featured us as one the top 15 new tech event tools; and CNET voted us a top start up at NY Tech Day.

"The Making Of" Story:
A recent transplant to New York, it was Maria's turn to organize her WINY Group. Frustrated with the email spamming and Facebook posts, she scoured the web for an easy way to create a good-looking mobile application. When she couldn't find one, she created Yapp.

Yapp Inc

Website:
yapp.us

Phone:
(347) 470-YAPP

Tips and Secrets:

Focus on listening to your users and providing amazing customer support. Sometimes it's the small things that delight your users and win them over and they then tell others about your app.

Marketing Techniques:

So far, Yapp's marketing strategy has been thoroughly organic. Word spread through our beta period by way of online media outlets and our beta launch was received with another boom in media coverage. But truly our app creators are our best sales people. When they create beautiful looking apps and share them with their communities, others get Yapp envy and come create apps of their own.

Risks and Challenges:

Like with any early stage venture, the risks and challenges are many from raising capital to building a product that people will use and love to finding others to leave their jobs and join your journey.

Setbacks:

At first, we didn't think that Android was that important of a platform. The first version of the product was native iOS and mobile web on Android. Users wrote to us that Android was important for them to feel the app was well distributed so we quickly re-prioritized and added a native Android version of YappBox. Another misstep was Facebook Login. We use Facebook as a quick account creation system. Many potential users were turned off by this as they no longer trust Facebook to keep their information private. We are now working on our own login system as an alternative.

Kahnoodle

FREE
By **Kahnoodle Inc.**

Description
Kahnoodle helps you shake up your relationship routine by challenging you and your partner to keep your love tank full.

Use Kahnoodle to:

- Create love coupons to show your love by fulfilling a romantic fantasy or completing a domestic task
- Record the sweet things you do for each other with "kudos"
- Discover the best offers available on date night activities in your city
- Chat with your partner when you are long distance
- Get tips and reminders for special things to do to express your love
- Create a wish list for places and things you'd like to do together

...All this and more in this fun & powerful relationship app!

Kahnoodle Inc.

Website:
www.kahnoodle.com
Phone:
202-657-6271
Email:
hello@kahnoodle.com

Profile:
About Kahnoodle's Founder Zuhairah Scott Washington:

Zuhairah is passionate about leveraging technology and digital media to build profitable, innovative, and useful products that improve human well-being and have a positive impact on society. She is an entrepreneur and seasoned business professional with over a decade of digital media, finance, and product/business development experience in global markets including Dubai, London, Los Angeles, New York, and Washington, DC working with Fortune 500 companies, vc-backed startups, and government and non-profit organizations.

Zuhairah is the Founder & CEO of Kahnoodle, a startup that aspires to be the biggest thing to happen to relationships since the advent of online dating. Through Kahnoodle, Zuhairah develops mobile products that help busy couples win at love – challenging the assumption that all technology is bad for relationships. Previously, she was the Director of Business Development and reported directly to the CEO and COO of Europe's largest local search and user-generated review website with 17M UUs in 10 countries and 7 languages. She also was a Principal & Co-Founder of Be Media LLC, an interdisciplinary, digital media consulting practice that partnered with clients to identify, launch, and grow products and services at the intersection of social impact and market opportunity. Prior to that she served as the youngest Regional Vice-President at a high-growth, entrepreneurial real estate private equity firm with $20B in AUM.

Zuhairah is acutely aware that the success and opportunities she has been afforded are not accessible to all. Consequently, she developed an online financial empowerment platform for women called The Billion Dollar Girls Club, which aims to equip 1 million low-income and minority young women and girls with the tools they need to increase their net worth and savings by $1,000 on average. Zuhairah graduated magna cum laude from UCLA and has a JD/MBA from Harvard. She has been singing and acting since the age of five and enjoys writing and recording original music in her free time. She is conversational in Spanish and has a soft spot for fine wine and all things chocolate.

Awards:

Noted by Entrepreneur Magazine as one of 100 Brilliant companies of 2012, Awarded Prize at Distilled Intelligence Startup Competition, Finalist, TechStars NYC.

"The Making Of" Story:

Zuhairah Scott Washington developed the concept behind KAHNOODLE as a newlywed when she became frustrated that the likelihood of being successful at her marriage was no better than flipping a coin and getting tails. She couldn't understand why there were online tools that allowed you to achieve other life goals like losing weight or saving money but there was nothing out there to help couples ensure their relationship success. Like any other forward thinking woman who sees a problem that needs fixing, she decided to build her own tool that could help other committed couples more proactively manage their relationship success.

Successes:

Over 5,000 downloads in less than four months. 300% month over month increase in downloads from Dec to Jan.

Tips and Secrets:

Build something innovative. Don't copy others or go for the lowest hanging fruit. Truly innovate – even if it is scary or different.

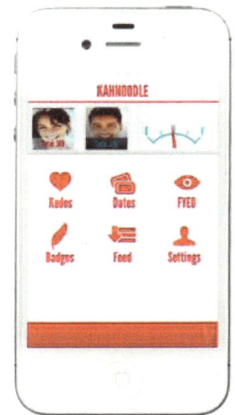

Marketing Techniques:

We started marketing by first writing blog posts for other sites looking for content. We also started speaking at conferences with key influencers. That allowed us to become known in the space and to be more credible – lending to more press.

Risks and Challenges:

We are a first time developer. We invested our own money into the project. This has never been done before but we knew it was a good idea that had been validated in other mediums – books, etc. We did a lot of customer development before building anything to make sure that we could deliver something that would give the user an AHA moment experience.

Setbacks:

We are constantly learning. We first started off with a gamified approach that made each partner compete against each other and that was more points based. We changed this and decided on a more collaborative team approach which felt more natural given the nature of most relationships. We are still learning and growing.

Rounds Video Chat Hangout Network

FREE
By **Rounds Entertainment Ltd**

Description
The popular Rounds comes out from Facebook to your mobile... A hangout like never before. Share pictures and watch YouTube together LIVE! Fun! Together! Now!

Hang out and make FREE video chats and FREE video calls with your friends! Capture moments, play with fun video effects, watch YouTube videos together and scribble funny images on each other. See what your friends are doing right now on Rounds through the photo stream, where you can like, follow and comment on all their great moments. Go on, show them some love!

Use Rounds Video Chat Hangout to connect with your family and friends.

Rounds Entertainment

Website:
www.rounds.com

Email:
support@rounds.com

Company Profile:
As the web's first social "hangout network," Rounds combines online entertainment with video communication to bring friends and like-minded people closer together for a fun, live experience across social networks, operating systems and devices. Bridging the offline and online worlds, Rounds uses shared activities, games and video to give friends the feeling of hanging out in real life. An interactive stream of friends' and featured users' "captured moments" from live sessions gives hangouts an afterlife, creating a new social network around them. Rounds has raised a total of $5.5 million in funding from industry leading investors, including Verizon Ventures, Rhodium, DFJ's Tim Draper, and other private investors.

"The Making Of" Story:

Our original concept was a speed-dating web app called *6Rounds*, conceived in 2008. Anyone who's been on a speed-dating experience knows they can be filled with plenty of awkward silences. Our idea was to provide real-time activities and games as icebreakers for would-be couples. That way, if there was no romantic chemistry but they still had five minutes to kill, at least they would have something to do together. During the development process, however, we realized our visionary concept of hanging out 'around' a shared activity could have many uses outside of speed dating. So, when *6Rounds* launched in July 2009, the speed-dating focus was gone, and we were a fun, new kind of real-time communication for everyone.

In October 2009, we were chosen to be one of Google Wave's first six extensions – its only video chat extension, in fact. Then, in August 2010 we launched our popular Facebook app, *Video Chat Rounds*, which helped us break through to younger users hungry for a more entertainment-based form of video communication. Throughout our whole history, we've never stopped adding new games, activities and features; to date, Rounds has over 7 million users worldwide.

Now with the launch of our iOS and Android apps, for the first time Rounds users will be able to hang out together and use many of Rounds' most popular features – including live YouTube video viewing together and applying fun effects on top of each other's video streams – directly from their mobile devices (both iOS and Android).

Successes:
Upon launching, Round Video Chat Hangout Network ranked #31 among all free apps in the US iTunes app store, and #4 for social networking (higher than Skype and Twitter). 40% of all mobile hangout conversations are "meaningful conversations" (3 minutes or longer). Rounds' average time per meaningful conversation is 11 minutes.

Tips and Secrets:
We started with a great product – built from the ground up to be viral. It's not only a great product that people want to share, we give them the tools <u>at the right moments</u> to enable that sharing. Other tips that were part of our strategy and which we recommend are:

- Knowing your target market; having a real solution for their needs; always marketing to that target
- PR – getting the word out there; not only getting coverage, but positioning your app in right way
- Cross promotion is very important if you have other apps you can use for that
- Having retention goals to keep users you've gotten
- SEO – choosing the right app name and description
- Doing usability testing, having strong customer support
- Localization – especially in-app; Rounds had its app store descriptions translated into 7 languages
- Getting featured by the app store helps incredibly

Partnership Opportunities:
Yes, Rounds is looking for developers and content partners.

Risks and Challenges:
Adapting the Rounds experience for mobile devices did have unique challenges. One of them was the need to streamline our features and user interface for mobile. Having *Rounds Video Chat Hangout* load and respond to users fast and reliably – without looking too cluttered for smaller screen devices was important to us. Another challenge lies in the way the mobile marketplace is structured. While the web is an open space where Rounds is free to do almost anything, mobile applications require approval from Google Play and the iTunes App Store for users to reach users. Programming for mobile required us to master new coding languages, and in the case

of Android, enable our app to look good on many different devices with unique specs and screen sizes.

We were able to capitalize on our successes from the web version of Rounds to test our ideas before implementing them on mobile. Having a large user base of 6 million users expedited that testing for us, and also meant *Rounds Video Chat Hangout* arrived in the marketplace with a fan base already interested in using it. Another advantage to operating in the mobile video communication space is that it's relatively uncrowded (compared to desktop-based options). Because Rounds had a clear vision from conception to make video communication a fun experience – and three years of experience implementing it – we're able to stand-out from the few other offerings, feeling much more like a richer 'hangout' experience for users, rather than a less dimensional 'video chat' one. And the hardware is ready for us; today's mobile devices are powerful enough to handle video communication properly, and front facing cameras are standard on most smartphones.

Setbacks:
The biggest hurdle we've experienced so far has been overcoming the famous chicken and egg problem with regards to users. We're a communication and entertainment platform that requires users to have at least one friend also using Rounds in order to enjoy what we have to offer. We solved this problem through a combination of very targeted marketing, developing Rounds in a way that offers users something to do even if they are the first to try it, and making sure it's easy to invite friends to join.

Touch™

FREE
By **Enflick**

Description
Touch redefines how we stay in touch with close friends and family.

Touch lets you:
Message your friends through Chats that feel like you are talking in person. Enjoy chatting with unlimited text, video, and voice messaging, delivery and read receipts, typing status, and top-notch speed and reliability.

Share your life with those closest to you through experiences that make the distance fade away.

Touch has been built to provide a personal, simple, yet powerful experience to let you maintain your closest relationships. Keep in touch with your friends on all major mobile platforms through Touch.

Company Profile:
Enflick was founded in the fall of 2009 with a mission to deliver innovative solutions that empower people to stay closer with their friends and family. Touch (designed to make staying connected with close friends and family an amazing experience) and TextNow (provides fast and reliable text messaging — for free) are the company's two core products. Both products are used by millions of users worldwide and are growing rapidly.

Enflick

Website:
www.enflick.com
Phone:
(226) 476-1578
Email:
derek@enflick.com

Awards:

- Derek Ting was recently named a finalist for the Ernst&Young Entrepreneur of the Year 2012 Awards.
- In 2012, Enflick made the list of The Achievers 50 Most Engaged Workplaces™ in Canada.
- nextMEDIA, in association with Deloitte, announced Enflick as one of 5 winners of the Top Canadian Digital Company category in the 2011 Digi Awards.
- In 2011, Enflick was named as a winner of Deloitte's Companies-to-Watch award
- In 2010, TextNow won the '2010 Best Social Networking App' in the Best App Ever Awards.

"The Making Of" Story:

We already had a user base of more than 21 million when we launched Touch, so we did have considerable input from customers, and we've always made listening to users a top priority. We saw that the mobile social networking market was exploding – the market data was pretty clear on this; however, the solutions that were out there focused on supporting very large numbers of acquaintances or specialized interest groups. Customers were telling us that they wanted a solution that gave them instantaneous social connectivity – *but* focused on the smaller circle of people they

wanted to keep in touch with the most. We heard it again and again from users. People wanted a solution to instantly connect with their innermost circle of friends and family, anytime and from any device. So in the case of identifying the opportunity for Touch, we heard it directly from our users and also experienced it ourselves from our own usage.

Successes:

Since its introduction in 2011, Touch has now served 800 million one-on-one messages and 150 million group messages. Since March 2012, the number of group messages has more than tripled, validating people's desire to more easily and quickly connect with their most significant relationships.

Tips and Secrets:

It always inspires me to watch people pursue their personal passions. Being exposed to fresh ideas – even if they're not related to our particular market – also gives a different perspective and helps the creative juices. Staying hungry to solve real user connectivity problems is critical to keep the momentum going as well. We're constantly testing our ideas and expanding if and when ideas are validated.

Marketing Techniques:

We use open forums for feedback, as well as social media tools and are always looking for opportunities to maintain a two-way dialogue with users regarding ongoing needs and wants. We're always paying attention to the things around us that we can improve and encouraging and listening to users' feedback. Also, our strategy includes a very flexible monetization model. Although we use an advertising supported revenue model, we also offer an option for users that don't want that and offer multiple levels. All of our pricing is competitive, offering the most functionality at the lowest possible price – and also free, of course.

TextNow + Voice

Free Texting and Calling

FREE
By **Enflick**

Description

FREE TEXTING, PICTURE MESSAGING, CALLING, AND VOICEMAIL

DEDICATED PHONE NUMBER
Give your friends your very own phone number!

UNLIMITED TEXT AND PICTURE MESSAGING
You can send as many texts and picture messages as you want - FREE!

TURN YOUR IPOD OR IPAD INTO A PHONE
Works on your iPod Touch, iPad, and iPhone

Enflick

Website:
www.enflick.com
Phone:
(226) 476-1578
Email:
derek@enflick.com

"The Making Of" Story:

We had a concept for a free text messaging product. We observed a huge demographic, particularly among young people, that wasn't being served by the existing texting model or 'plans' that were delivered by carriers. Being located near the University of Waterloo with a large student population, this need was especially apparent to us. We saw a market that was cash strapped yet had a huge desire and interest in connecting with others. Further, we knew that this market was also extremely tech savvy. We wanted to develop a connectivity solution that was easy to use yet highly accessible and affordable to this market – and that concept came to be TextNow.

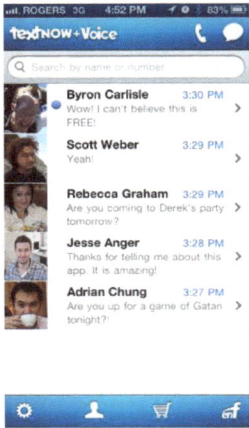

Successes:

With TextNow, Enflick's first product, Derek Ting established a multi-million dollar profitable business with no initial funding. Derek has led the development and market introduction of multiple and successful software connectivity products that today serve a network of more than 25 million users worldwide. In March of 2011, TextNow sent its billionth message.

Risks and Challenges:

Our biggest risk was probably TextNow – basically developing and delivering what we knew would be a free service. There's definitely some risk when you're investing personal resources and time and putting a solution out there for free! But doing so was integral to our mission – which was to make it accessible and affordable for everyone to send real-time messages to each other. We knew that obtaining momentum would be key for the ultimate success – and monetization – of TextNow.

Setbacks:

To gain momentum, we decided to start on the iPod Touch platform because it was highly available at that time to our target market. It made it really simple for users to try the product – they only needed an iPod Touch and a WiFi connection. As it was in 2009, the iPod Touch was 'the' big item, especially among young people, that year, so the timing added to our momentum as well. We also worked to get the support of app reviewers and bloggers – getting our product to them, getting them to use it, so they would publicize its value among their followers. Following this strategy really paid off, with TextNow going viral in December 2009. We had so much traffic coming in that we basically skipped the holidays with our families that year. We needed to be available to ensure the servers didn't crash due to the overwhelming number of users.

Topi

FREE
By Topi

Description
The new way to make connections!

Meeting relevant people is like looking for a needle in a haystack, particularly at huge conventions. Shouldn't there be an easier way? After all, it's 2013!

Topi allows you to quickly and safely connect with others nearby based on what matters to you! Browse people, chat in our topic-based rooms, invite your friends and start private group conversations, as you would do in real life.

Company Profile
Founded in 2011, Topi is the first mobile solution to promote and facilitate networking and interactions at conferences. Topi advances the notion of "serendipity by design" by serving up more relevant and meaningful information to people and significantly improving their professional networking experience. Make the most out of conferences! Quickly connect with like-minded professionals around common interests and backgrounds.

Topi

Website:
www.topi.com
Phone:
646-403-4335
Email:
david@topi.com

David Aubespin, Founder & CEO
A former Google engineering manager, David has over 14 years of experience managing technology teams and building products. At Google, David worked closely with top-tier customers of AdWords. Prior, he spent 2 years at IBM Watson Research Center working on Quality of Service for web servers. He also served 4 years as senior software engineer and technical architect at eMeta, a startup in New York that was

acquired by Macrovision in 2005. David holds a Master of Science in Computer Science from University of Nice, France and an MBA from Columbia Business School.

"The Making Of" Story:

The best-seller book "Never Eat Alone" prescribes how to prepare and make the most out of networking at any event, but some of us are hard-pressed for time and unable to do this much homework in advance. That was the inspiration for David Aubespin to build a new mobile app that would improve event networking for both attendees, as well as the conference and event organizers. Topi makes it easy to quickly connect with like-minded professionals around common interests and backgrounds during and well after events have ended.

With Topi, business professionals can engage with other attendees in new ways. A unique feature is its powerful "PeopleRank," which provides a searchable view of other attendees, weighed and sorted by relevance to the user, letting them quickly identify others most important to them to meet based on common affiliations, field of expertise, job titles, areas of interests, etc. Topi's technology adds a compelling layer of context by surfacing key topics to help kick-off conversations among mutually interested but new parties.

Attendees can use Topi to facilitate group or 1-on-1 conversations, public and private, during and after the event has ended. In addition to text, they can communicate with photos, maps, audio, sketches, etc.

At the same time, Topi provides conference and event planners with a self-service website to "mobilize" their events quickly. Within a matter of minutes, all event material, speakers' bios, agenda, etc. are accessible via the native iOS, Android and Windows Phone apps.

Once an event is created, organizers can leverage Topi's robust sharing and communication features: instant polling and surveys that can be pushed out on the fly or scheduled to attendees, real-time answers aggregated and shared, graphical analytics and demographics captured, and in-app notifications to send scheduling changes, conference materials, updates to attendees, etc.

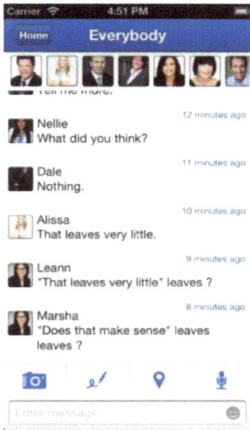

A unique feature is Topi's sophisticated geo-fencing. Event organizers map out with precision multiple locations where an event is happening (i.e., central venue, remote sites, etc.), and attendees within predefined perimeters can participate and socially interact with other attendees.

Topi is also seamlessly integrated with social networks including Facebook, Twitter, LinkedIn, Foursquare and Instagram, and the two largest online event platforms Eventbrite and Meetup.

David's vision for Topi put organizers and planners in command of mobilizing their events, with minimal time or effort, while raising the bar for attendees by improving their professional networking experience. With Topi, events leave a worthwhile, lasting impact with attendees, helping conference directors and meeting planners feel confident about the value and success of their events.

Successes:
Topi is used by a variety of organizations, such as American Express for its Marketing Mixer events in New York or Duke University for its Global 2013 event. Some interesting uses of Topi included the InfluencerCon event, which took place in five countries simultaneously. Topi was set up so that attendees in each country could see everyone as if in the same physical location. Children's hospitals have also started using Topi in order to connect children and help them recover more quickly, by being emotionally engaged with other children across the country. More recently, Topi has ventured into white labeling and developed a customized version for PlanYourMeetings to integrate elements of augmented reality at their events.

Tips and Secrets:
The main reason for which event organizers are amazed by Topi is because Topi did not simply make things better or easier to use, Topi reinvented networking at conferences. David and his team decided early on to work on what made sense to them, without focusing on what the status quo was. As a result, event planners can offer a very unique experience to attendees every time they use Topi.

Marketing Techniques:
Topi constantly keeps event planners, conference organizers, and convention industry bloggers in the loop when developing new features or launching new services.

Consequently, influential entities become very early adopters and pave the way for broader deployment.

The Topi team now also attends startup competitions, as the app is a natural extension for these events. Most recently, Topi was presenting in New York at the Projective Space Demo Night.

Risks and Challenges:

When David initially reached out to event organizers, a number of them were concerned about sharing attendees' data and proprietary material with an external provider, and preferred to build their own app instead. However, after David demonstrated how securely data could be shared and how only aggregated profile information was used by Topi to match attendees, most event organizers decided to use Topi for their events.

David also faced the challenge of building a mobile offering that could be made available to 85% of the smartphone population - iOS and Android users. However, building native apps that run cross-platform is a big challenge that is often too time-, cost- and resource-prohibitive for many. It takes many developers upwards of several months to build just one native app, say, for iOS and then another several more months to port that app to run another platform like Android. David built Topi as a native iOS, Android and Windows Phone app in only six months with only an angel investment.

Setbacks:

When Topi was first looking at coming to market, social discovery apps were all the rage in the early spring of 2012. Yet, the rage fizzled quickly and social discovery as an emerging technology faced an uphill battle in mainstream consumer adoption due to many privacy and security issues among other challenges. At that time, Topi could easily fit this broken mold and came out as a social discovery app, but David Aubespin was smart to recognize that problem very early on and quickly pivot his idea into something much greater — filling a need in the $263 billion conferences and meetings industry. David jumped in and built an innovative networking app for events, which smartly surfaces for business professionals - the right people to engage in richer conversations, the right opportunities to connect, and build more meaningful relationships.

Index

www.ingramcontent.com/pod-product-compliance
Lightning Source LLC
Chambersburg PA
CBHW041441210326
41599CB00004B/94